Introduction to the Primates

Introduction to the
PRIMATES

Daris R. Swindler

Illustrated by

Linda E. Curtis

University of Washington Press

Seattle and London

Library of Congress Cataloging-in-Publication Data

Swindler, Daris R.
Introduction to the primates / Daris R. Swindler ; illustrated by
Linda E. Curtis.
p.   cm.
Includes bibliographical references (p.   ) and index.
ISBN 0-295-97704-3 (alk. paper)
1. Primates.   I. Title.
QL737.P9S82   1998
599.8—dc21   97-47149
                                                            CIP

Chapter opening illustrations are unattributed early nineteenth-century bookplates
collected and purchased by the author in Cambridge, England.

To the memory of two friends and colleagues,

James A. Gavan and Elizabeth S. Watts

Primatology

Carefully preparing to speak,
raised and perched like monkeys in the canopy;
sharp, hairy hands,
described to the edges of sleeves
without significance
in unillustrated space.
A broadleaf
laced in the upper branches,
about to consider the next sequence.

The continuing movement,
an outline of air
that starts the arm elastic
in round reach downward,
one over another,
a swing in and out of detail

like sense, like strings,
connected place to place.
Insouciant.
Bending,
dropping by layers in leaf,
smoothly
drooping a branch with weight,
swooping
over, above ground,
defining distinctly
a jungle
with vines.

Kathryn Rantala
1996

ᔰ Contents

⌒ Foreword

In the last thirty years, research in the field of primatology has revealed to us a new world, the world of our closest mammalian relatives, the primates. This research has brought to our attention the wondrous diversity of primate adaptations to their environment, from the baboons of the East African savanna to the lemurs of Madagascar and the orangutans of the Bornean rain forest. This effort has also narrowed the gap between "us" and "them": The more we learn about primate behavior, the closer the primate family becomes—humans, apes, monkeys, and all. We as humans can no longer boast of being "primates, but different," because those once-vaunted major differences between humans and nonhumans—in cognition, communication, and tool-making ability—are looking increasingly subtle. We are separated from our primate relatives by differences of degrees, not kind.

I was introduced to the world of primate biology by Professor Swindler more than twenty years ago. At that time, when I was a first-year graduate student in the Department of Anthropology at the University of Washington, I enrolled in his course in Comparative Primate Anatomy. Because the course was a degree requirement, my intent was to get it out of the way early in my graduate career so that I could go on to focus more exclusively on my chosen areas of biochemistry and molecular evolution. Professor Swindler and his class changed my mind and my direction decisively within weeks. Before my eyes, the lives of primates present and past unfolded in vivid lectures and engaging labs. His classes were never dull events, not only because of the inherently interesting subject matter but also because of his unique way of putting a humorous twist on improbable details. I soon found that I was much more interested in discovering the intricacies

of primate history, as revealed through comparative anatomy and paleontology. Professor Swindler reached out to everyone in that class, from
committed graduate students to zookeepers to pre-med students, because his own knowledge embraced nearly every aspect of primate biology. This book allows you, the reader, to experience some of the
magic of that class and to be introduced to the scope of primatology as
seen through the eyes of one of the field's most renowned researchers
and educators.

This excellent introduction to the world of primates reflects Professor Swindler's particular interests in primate anatomy, growth, and development. And, just as in his lectures, there is something here for
everyone, from budding students of paleontology, ecology, conservation, and medicine, to students with just a healthy all-around curiosity
about their closest animal relatives.

As time passes and humans exert more influence over their physical
environment and the other animals in it, it is important for us not to
lose sight of our heritage as primates or our responsibilities to our fel-
low primates. This book will help you appreciate the grandeur of our
primate legacy by describing some of the inhabitants along the wind-
ing trail of forest-dwelling creatures from which our distant ape ancestor arose. It will allow you to appreciate how they evolved to look as
they do and how they interact with themselves and their environment.
Very importantly, it will also introduce you to the myriad species of primates that still make their homes in the world's diminishing tropical
forests and other threatened environments. These species, over which
we command enormous power, rely on us for continued survival. They
have no voice to raise in their own defense except that within each of
us. With every primate population or species lost, a part of our own
heritage disappears forever. Read this wonderful book with awe and appreciation, and with thanks to Professor Swindler and all our primate
relatives.

Nina G. Jablonski
California Academy of Sciences

◟ Preface

Primatology is a relatively new discipline even though its roots lie deep in the history of zoology. The intrepid Carthaginian explorer Hanno (fifth century B.C.) had magnificent, if somewhat exaggerated, stories to tell about his encounters with monkeys and apes along the west coast of Africa. Among the plethora of ancient Egyptian gods, the baboon was worshiped and venerated for thousands of years. Monkeys and apes were depicted in the writings, drawings, and sculpture of many ancient civilizations. In one manner or another humans have long been aware of and interested in monkeys and apes and their similarity to ourselves.

It was left to the English physician Edward Tyson to publish the first systematic treatise of a nonhuman primate in 1699. This was a milestone in the history of anatomy and zoology, and it remains as one of the great and lasting contributions to the science of comparative anatomy and what later was to become primatology. From this time on, monkeys and apes were looked upon with more understanding and compassion, and, what was equally important, their proper biological position within nature began to be appreciated by a much larger audience.

Some fifty years later, the great Swedish botanist Carolus Linnaeus began publishing his *Systema Naturae*, which by the tenth edition in 1758 included monkeys, apes, humans, and bats all in the order Primates. It should be noted that in later editions bats were removed from the order. A short one hundred years later, the English naturalist Charles Darwin published his *Origin of Species*, and *The Descent of Man* appeared two decades later. Between these landmark events, the quest for a clearer understanding of the primate's place in nature was adroitly and succinctly detailed by Thomas H. Huxley in his classic *Man's Place in*

Nature (1863). By the beginning of the twentieth century, few scholars doubted the biological kinship between humans and nonhuman primates. The investigations which span the last three-quarters of the twentieth century have merely confirmed the closeness of this natural bond.

The aim of this book is to introduce readers to a group of mammals known as primates, specifically nonhuman primates, who along with ourselves are classified in the order Primates. This book is meant as a beginning resource. For many, it will be complete in itself; for others, it will provide the background necessary for further studies. For all readers, it is hoped that the material will be interesting.

While a lot of detail and comparison is offered in these pages, this book is intended both to satisfy and kindle the reader's interest, and hopefully will provide a broader picture of a topic very close to us. The information in this book is not complete, because our knowledge of the subject is not yet complete. The relationship of prosimians, monkeys, and apes to ourselves—and our relationship to the entire animal kingdom—is a fascinating and still unfolding story.

Daris R. Swindler
Edmonds, Washington

⌁ Acknowledgments

Many people have contributed to the writing of this book, some directly by reading portions of it and offering comments and suggestions, some more indirectly through conversations. I wish to thank, in particular, Nina G. Jablonski of the California Academy of Sciences and Mark Terry of the Northwest School (Seattle), who read an early draft of the manuscript and made many valuable suggestions. Thanks too are accorded to Carolyn M. Crockett of the Regional Primate Research Center (University of Washington) for her thoughtful and useful comments and suggestions. I also wish to thank the reviewers for their insightful comments and suggestions.

Lee Haines (University of Washington) and Edy Schmoker (Seattle) deserve special thanks for their excellent photographs. Here I would also like to thank my colleague Linda Duchin (Tacoma Community College and the University of Washington) for her support and encouragement during the preparation of the book and Jackie Pritchard and the staff at the Primate Information Center for able assistance with source material.

My wife, Kathy Swindler, has been closely associated with the writing of this book from its conception. Her patience and understanding, as well as her many helpful suggestions, are gratefully acknowledged.

I would like to extend a special acknowledgment to the illustrator, Linda Curtis, for her diligence and for the quality she brought to the original illustrations throughout the book. The illustrations at the beginning of each chapter are unattributed eighteenth- and nineteenth-century bookplates found in Cambridge, England.

And, finally, I wish to thank Naomi B. Pascal and the staff of the University of Washington Press for their help.

Introduction to the Primates

1

Monkeys
and Apes
in History

Humans have a special interest in the group of animals known as primates since, in addition to prosimians, monkeys, and apes, it includes ourselves. This interest is traceable into remote antiquity, long before it was realized that we humans are a part of nature and that primates are our closest relatives. Inquisitiveness is a legacy of those primate origins. Who among us has not wondered about those familiar faces peering at us from the monkey island at the local zoo. How could this smiling face possibly remind us of cousin Doyle or aunt Helen? As we shall see, these and similar questions were undoubtedly on the minds of many peoples living in and around the regions inhabited by monkeys and apes.

Monkeys were probably known long before apes. One of the earliest paintings of monkeys dates from the Old Kingdom of Egypt (2575–2134 B.C.) and depicts a hamadryas baboon and other monkeys among the animals on the wall of a tomb at Thebes. Baboons were highly regarded by the ancient Egyptians who associated them with Thoth, one of the gods of the dead and author of the *Book of the Dead*. Thoth was responsible for recording the verdict on the soul, and was shown either as an ibis or a baboon (Hobson 1987). By the time of the New Kingdom (1550–1070 B.C.), great numbers of animals were being mummified. At one town, Tuna-el-Gebel, a cemetery dedicated to Thoth contained thousands of mummified baboons and ibises (El Mahdy 1989). The statue shown in figure 1.1, probably a votive offering, depicts a hamadryas baboon carved from lapis lazuli, a semiprecious stone common in Egypt.

Although the Egyptians never really worshiped baboons, they certainly venerated them and used them and other animals to depict the gods associated with them. Baboons were also household pets and were often trained to climb fig and date trees to pick the fruit. There are many wall paintings showing baboons on a tether sitting atop a date palm tree throwing dates down to the owner.

At the ancient city of Mohenjo-Daro in the Indus Valley of India (2500–1500 B.C.), archaeologists found a terra-cotta figure that may represent a rhesus monkey (Reynolds 1967). Certainly rhesus monkeys are common in this region today. This and other figurines found in Mohenjo-Daro are similar in workmanship to those from Egypt, although there is no positive evidence that the two regions were in contact with each other. For hundreds of years the sacred langur, or hanuman monkey, has been dedicated to Hanuman, the famous monkey-god of southern India, and has had a special meaning for thousands of devout Hindus.

One of the more famous and certainly better known Oriental monkey representations is the group of three sacred monkeys of Japan. These are the three "wise monkeys" posed to represent the three primary dicta of the Buddhist religion: one with ears covered, a second with mouth covered, and a third with eyes covered, which simply means "hear, speak, and see no evil of anyone or anything."

"Monkeys," a comedian once said, "are the craziest people." We cannot be absolutely sure what the comedian had in mind, but it is true that through the millennia at least the predominantly Christian regions of the world regarded the monkey as a sort of clown: somewhat foolish, devilish, often greedy, and perhaps even a little silly. "Apish" is a term often used to describe the behavior of either monkeys or apes. In *Webster's New Collegiate Dictionary* the word has three definitions, of which two are interesting and revealing regarding a common attitude toward the behavior of monkeys and apes. The first simply means "to resemble an ape." The second and third, however, refer to slavish imitation and behavior that is extremely silly or affected.

1.1 An Egyptian lapis lazuli statue of a hamadryas baboon. Ancient Egyptians often mummified baboons, which they associated with their god Thoth. (Author's collection)

When and where did these ideas get started? An early Hebrew text says that "when Noah tasted the first wine, the Devil sacrificed a sheep, a lion, a monkey, and a pig, in that order. These symbolized the stages of man in drunkenness—mild, wild, foolish, and finally disgusting" (Reynolds 1967, pp. 27–28). For whatever reason, the monkey has long been looked upon as a buffoon or devilish figure and has maintained this characterization through the ages, particularly in Christian countries. The perception of the monkey as an incarnation of the devil is well illustrated by a painting of an unknown European artist of the eighteenth century (fig. 1.2). In time monkeys came to be looked on as rather disgusting, bestial creatures whose describers were more concerned with their moral attitudes than with reporting the habits of the animals (Richard 1985).

The ancient Greek poets, and many of Aesop's fables, related stories and anecdotes about monkeys being easily tricked or duped by other animals and often represented them as symbols of ugliness. Certainly the Greeks were familiar with several types of monkeys by the fourth

1.2 Drawing of a monkey by an unknown European artist of the late eighteenth century.

century B.C., since Aristotle (384–322 B.C.) gave us what may well be the first primate classification in his *Natural History of Animals*:

1. Monkeys with no tails (Barbary apes)
2. Monkeys with tails
3. Dogheaded baboons

The source of monkeys in early Greece as well as in Etruscan Italy, where monkey paintings have been found, was likely the great North African port city of Carthage, whose ships had been plying the Mediterranean Sea for several centuries by this time. A sea captain of one of these vessels was a Carthaginian by the name of Hanno, an intrepid explorer who in 470 B.C. set sail with sixty five-oared vessels to colonize and trade along the west coast of Africa. According to most authorities, Hanno probably sailed as far south as the present coast of Sierra Leone where, upon landing on an island in the bay, he encountered several "hairy sylvan creatures who replied to the seafarers by throwing stones at them" (Hartmann 1901, p. 2). The sailors killed three of these creatures that the local natives called "gorillas." They were skinned and brought to Carthage, where at least two of them remained, according to Pliny the Elder, until the sack of Carthage by the Romans in 146 B.C. (Hartmann 1901).

We know today that Hanno did not get as far south as the present distribution of gorillas, and therefore these creatures were more likely chimpanzees, which is even more probable since they were throwing stones and using sticks to ward off the attacking sailors, a behavior not known to gorillas. As noted by Schultz (1972, p. 2), Hanno did supply solid evidence for the then "popular belief in beings which seemed to be neither man nor animal." In other words, these skins reenforced the common belief in satyrs, fauns, sphinxes, and other mythical creatures resembling humans.

Some naturalists today believe that satyrs, one such interesting mythological creature, may actually have been gibbons: others believe that we cannot be sure of this connection. The satyrs of Greek mythology were forest dwellers, half goat, half human, who supposedly fled to the protection of mountains when disturbed. Reynolds (1967) stated that according to Aelian, a third-century A.D. writer, satyrs also inhabited India. Interestingly, there is one group of gibbons still living in eastern India and Bangladesh today.

In the second century A.D. one of the most famous Greek physicians and anatomists undertook the dissection of Barbary apes (they are not apes but a tailless monkey living in North Africa), baboons, and perhaps other types of monkeys as well. This was Galen (A.D. 131–201?), one of the greatest early anatomists, often considered the last great biologist of antiquity. He wrote sixty-two treatises on the form and function of the human body, though it is doubtful that he ever dissected a human body at all. His writings, though based on nonhuman primate anatomy, were accepted by the Christian church, and for the next twelve centuries they were considered infallible and constituted the only text on human anatomy available to medical schools in the Western world.

For the next thousand years there was little progress in science, and the so-called Dark Ages were in full swing. Toward the end of this period, in the eleventh and twelfth centuries, trade began to increase between distant parts of the world, and ships brought many strange animals, including monkeys, into Europe. Ideas were beginning to change, and a more worldly understanding and awakening was slowly emerging in Europe, although it was to be several centuries more before the great sea powers of Europe began the explorations that were to change forever the human attitude toward the world and its inhabitants.

Some of these early explorers are familiar to us—for example, Marco Polo (1254–1323), Vasco da Gama (1469?–1524), Christopher Columbus (1446?–1506), and Ferdinand Magellan (1480?–1521). These explorers, and the intellectual changes that occurred after them, set the stage for a whole new interest in and understanding of nature, animals, and their relationship to humankind. Magellan was the first explorer to circumnavigate the world, thus disproving the ancient notion of a flat world.

At the same time that these intrepid seafarers were exploring the world, a new birth of science, art, and philosophy was slowly emerging in Europe after centuries of the virtual stagnation of intellectual curiosity. The Renaissance had arrived. The printing press, invented in the middle of the fifteenth century, was to prove one of the most versatile and efficient methods of disseminating manuscripts and books ever invented. Scholars could now, in a relatively short time, read and compare their information with that of scholars from other parts of the world. Printing may well have been the single most important aid in the revival of learning.

The sixteenth century was the end of the so-called Middle Ages and the beginning of the Renaissance of Science. This was primarily due to the publication of two books, one by a Catholic priest, Nicolaus Copernicus (1473–1543), on the revolutions of the heavenly bodies. After years of careful observations, Copernicus was able to present convincing evidence that the sun was the center of the universe, breaking with the long-held Ptolemaic interpretation that the earth was the center. The other book, also based on observations (of the human body by meticulous dissection), led to the renunciation of the Galenic dogma concerning the structure of the human body. The author of this book was a young professor of anatomy at the University of Padua, Italy, by the name of Andreas Vesalius (1514–1564). The book was simply titled *De Humani Corporis Fabrica*, or "On the Structure of the Human Body." These two books, published within a week of each other and based on the method of observation, were destined to forever change the human attitude concerning the universe and nature, not just because of their content but, perhaps more importantly, because they relied upon "induction, the method which, in a phrase, resumes the processes by which from well-observed particular instances general principles or laws may be derived. . . . [It was] in short, the method of science" (Ashley-Montagu, 1943, p. 12). *Fabrica* thus became the model for the development of the science of the anatomy of other animals compared with humans.

About this same time Konrad Gesner (1516–1565) published *Historiae Animalium*. This work eventually comprised five volumes, which are considered to be the first illustrated books on zoology. Gesner described several kinds of monkeys and apes. He did not dissect or study these animals, rather his descriptions were taken from other sources and he frequently relied on "travelers' tales," a common practice in those days. Gesner is generally thought of as the founder of the modern science of zoology.

In 1641, a Dutch physician and anatomist, Nicolaas Tulp (1593–1674), described what he called *Homo sylvestris, Orang-outang*. According to Ashley-Montagu (1943), this was the first use of the name "orangutan" by a European writer. The name was in common use at this time for any manlike ape that was not a monkey. It is a Malayan term and means "man of the woods" or, in Latin, *homo sylvestris*, and continued to be used into the middle of the nineteenth century for all great apes. In fact, when Savage and Wyman published the first scientific account of the gorilla in 1847 they referred to it as a "new species of Orang" (Ashley-Montagu 1943, p. 249). We will shortly see that this was also the name used by Tyson for the animal he studied in 1699.

Tulp's ape came from Angola, on the west coast of Africa, and was either a chimpanzee or a gorilla—most probably, it is now thought, a chimpanzee. Reynolds (1967) believes that it was probably a pygmy chimpanzee due to its size and the fact that it possessed webbing between the second and third toes, a feature frequently found in pygmy chimpanzees today.

Ashley-Montagu (1943, p. 250) presents an English translation of a part of Tulp's text regarding the behavior of "Orangs," undoubtedly taken from the ship's captain or crew that had brought the animal from Angola. From the description, it seems to have been based largely on the comments of the local natives. It reads as follows: "In fact they are so greatly inclined to venery (even among themselves, as was common with the licentious Satyrs of the ancients) that they are at all times wanton and lustful: so that the Indian women therefore avoid the woods and forests, worse than dog and serpent, where these shameless animals roam." Unfortunately, it was such promulgations, taken from the tales of explorers and based on native stories, not personal observations, but accepted nonetheless as truth, that perpetuated the view of monkeys and apes as indolent and lascivious creatures well into the nineteenth century.

Edward Tyson (1650–1708) is credited with the founding of the modern study of primatology with the publication of *Orang-Outang, Sive Homo sylvestris, or, the Anatomy of a Pygmie Compared with that of a Monkey, an Ape, and a Man* in 1699. It is interesting to note, however, that according to Schultz (1971) the term primatology was not used in print until 1941 in the famous *Bibliographia Primatologica* by Theodore Ruch.

Tyson was born in Bristol, England, studied medicine, and was a distinguished physician; but much more, he was the leading anatomist and anthropologist of his day. He was a complete naturalist. He not only taught human anatomy but also carried out detailed, accurate investigations of the anatomy of the porpoise, tapeworm, opossum, and chimpanzee and held what many believe to be ingenious ideas regarding evolution, anthropology, and especially comparative anatomy. In his excellent biography of the life and times of Tyson, Ashley-Montagu (1943, p. 387) writes that Tyson was not only the foremost comparative anatomist in England during his life but was also "the greatest of England's comparative anatomists, and among the greatest of all time anywhere." The methods Tyson developed and applied—that is, describing each feature and comparing them feature by feature during his dissections of the chimpanzee, monkey, and man—permitted inferences and conclusions about structural relationships and function. Here was the inductive method again: conclusions being gleaned from observations of particular facts.

Although not an evolutionist in the sense in which the term is used today, Tyson did believe that the various groups of animals were related one to the other. He coined the term "chain of creation" when discussing the relationships between different groups of animals, and as Reynolds (1967) correctly points out, this was a gradational scheme rather than an evolutionary one, one that may hark back to Aristotle's *scala naturae*, or the scale of nature. Tyson is one of the most distinguished comparative anatomists of all time and may be justifiably considered the founder of both comparative anatomy and primatology.

One year before Tyson died in 1708, Carolus Linnaeus (1707–1778) was born in Sweden. Linnaeus was a professor of botany at the University of Uppsala and had a penchant for classifying plants and later animals. He was only twenty-eight when he published the first volume of *Systema Naturae* in 1735. By then he had worked out definitions of the terms "genus" and "species" and the methods of classifying animals within them, a method which still, with some modifications, forms the basis of the system of classification used today (classification is considered more extensively in chap. 2). Linnaeus, as was true of many of his predecessors, was not an evolutionist but believed in the fixity of species, that is, independent creation with no changes occurring after the moment of creation. The belief that species were immutable was common at this time, and although Linnaeus weakened in this opinion somewhat as he got older, he always maintained that species were fixed units. The tenth edition of *Systema* in 1758 is generally considered to be the beginning of the modern system of classification, for it was in this volume that he first employed the binomial system still used today.

In this binomial system all plants and animals are given two Latin or latinized names, of which the first (genus name) is a group name which may be shared by several species, while the second (trivial name) is particular and identifies the individual species. The genus name is capitalized, the trivial name is not. Thus, *Gorilla gorilla* refers to all gorillas living in Africa today. But when a species is further divided into subspecies the latter are designated by trinomials: for example, *Gorilla gorilla gorilla* refers to the lowland gorilla, while *Gorilla gorilla beringei* is the mountain gorilla.

It was in this famous tenth edition that Linnaeus first used the term primates (the "first" or "highest") to include four genera: *Homo* (man), *Simia* (all monkeys and apes of the Old World and New World then known), *Lemur* (lemurs as well as the "flying lemur," which we now know is not a primate), and *Verspetilio* (bats). Bats were later removed from the order Primates. Linnaeus recognized the close similarity among monkeys, apes, and humans (and, he thought, bats) by putting them in the same group, which he defined as having four parallel upper teeth (incisors), two mammary glands on the chest, and fully formed clavicles. The next step, however, to suggest that this similarity indicated a biological or genetic relationship, was never proposed by Linnaeus. In fact, a hundred years later, even Darwin sidestepped this issue in *The Origin of Species* (1859) for he wrote only one short sentence in the entire volume on the subject of human evolution: "Light will be thrown on the origin of man and his history" (p. 488).

Many scholars of the day disagreed with Linnaeus's placement of monkeys, apes, and humans in the same biological group, and several other classification systems appeared within a few years of Linnaeus's that attested to the many anatomical differences between monkeys and apes on the one hand, and humans on the other. The one physical, unique attribute of man among primates that all authorities of the time emphasized was that man had only two hands instead of four. This position was supported by the early student of anthropology Johann Blumenbach (1752–1840), who proposed two divisions, Bimanus (man only) and Quadrumana (lemurs, monkeys, and apes). Usage has shifted to refer to the feet rather than the hands, and now we would say that humans are the only bipedal primate and all other primates are quadrupedal.

The brilliant French naturalist Georges Louis Count de Buffon (1707–1788) was born in the same year as Linnaeus. He made many contributions to comparative anatomy and zoology, writing twenty volumes of natural history on subjects ranging from cosmogony to insects. In these volumes he also discussed the mutability of species in relation to changes in environment and thus differed from Linnaeus's idea of the fixity of species. Buffon was the first naturalist to use the term "gibbon" for an ape he described in 1785 (Reynolds 1967).

During the eighteenth century the relationships among the apes were being sorted out by various scholars. This was a slow process since only a few people had ever seen these creatures dead or alive. At about the time Buffon was describing the gibbon, the great Dutch anatomist Pieter Camper (1722–1789) was dissecting several orangutans he had procured from Dutch shipping companies. It was these dissections which, when he compared them with Tyson's studies, led him to conclude that the chimpanzee and orang were different animals living in very different parts of the world; the chimpanzee in Africa, the orang in the Indies.

It was not until almost the middle of the nineteenth century that the first gorilla was described scientifically (Savage and Wyman 1847). Until the nineteenth century the gorilla was known only from travelers' tales and hearsay, mostly anecdotal. Dr. Savage was an American missionary, and Dr. Wyman, an American anatomist. They published their account in the Boston Journal of Natural History, and called their specimen Troglodytes gorilla, believing it to be a new species of orang. The genus name was later changed to Gorilla, which is now also its common name. People were curious about this giant ape, a creature that had been known from folklore for centuries, and now there was finally positive proof of its existence. Unfortunately, the gorilla received bad press regarding its habits and especially its behavior toward people. Dr. Savage himself wrote of the incredible strength and indescribable ferocity of the gorilla and how dangerous it was to hunt. An often-quoted passage from one of Savage's essays bears repeating here since it so clearly describes what was then believed to represent gorilla behavior: "When confronted with a gorilla the hunter must stand his ground while listening to the horrifying cries and watching the onrushing monster, and then as the animal gets closer and closer, the courageous hunter steps forward and puts the barrel of the gun in the gorilla's mouth and pulls the trigger. If the gun fires the gorilla drops dead but if the gun does not fire, the gorilla crushes the barrel between its teeth reversing the intended order of events" (1847, p. 423).

The intrepid American explorer Paul du Chaillu left his homeland in 1855 and spent the next four years exploring in Dutch equatorial Africa before returning to America. He wrote an account of his adventures, *Explorations and Adventures in Equatorial Africa*, which appeared in 1861. This book caused much controversy and a great deal of bitterness between du Chaillu and many of his contemporaries, some of whom doubted the authenticity of many of his claims regarding the customs of local natives, as well as his observations of animal habits. He decided on a second African expedition and spent the years between 1863 and 1865 again in equatorial Africa. There is little doubt that he explored many unmapped areas of western equatorial Africa, making large collections of plants, animals, and native materials. He published several other books about his travels and always claimed to have been the first white man to have killed a gorilla, which he described as the King of the African forest (du Chaillu 1890; fig. 1.4).

Du Chaillu was certainly a flamboyant man, given to hyperbole in his books and public lectures, prone to antagonizing many with whom he came in contact during those years. He was not the foremost naturalist of his day and did made mistakes in identification of new species of apes, but, to do him credit, he did not accept several then current beliefs regarding gorilla behavior. For example, he wrote that gorillas did not lurk in trees to "pull-up unsuspecting passers-by and choke them to death and later eat them." He said, "They are ferocious, mischievous, but not carnivorous" (du Chaillu 1890, p. 226). He had studied the contents of the stomachs of many of the gorillas he shot and never found anything except leaves, berries, and vegetables. He did not believe that gorillas attacked elephants and beat them to death with sticks, or that they carried off women from the native villages, one of the most repeated tales concerning gorilla behavior for many years. Today we

1.3 Hunter killed by a gorilla. (P. Du Chaillu 1890)

1.4 "My first gorilla." (P. Du Chaillu 1890)

know that the gorilla is one of the most pacific of animals with regard to humans.

Du Chaillu had encountered what we know today as the lowland gorilla, *Gorilla gorilla gorilla*. He also discovered and described what he considered to be a gorillalike chimpanzee he named "kooloo-kamba." The existence of this enigmatic creature has been debated for over a century, and as recently as the 1960s the late primatologist W. C. O. Hill (1972) placed it in a subspecies of chimpanzee, *Pan troglodytes kooloo-kamba*. It is doubtful if such hybrids actually exist; it seems more likely that they represent extremes of development within chimpanzees and gorillas. For as Shea (1984, p. 11) suggests, "This is probably why it is large and robust chimpanzees (or small female gorillas) which have been labeled kooloo-kambas or hybrid forms." The importance of differences in growth and development is more thoroughly discussed in chapter 8.

The famous mountain gorilla, *Gorilla gorilla beringei*, was discovered in 1902 by Oskar von Beringe in the volcanic mountains of central Africa in what today is Rwanda and Uganda. It was thought at first that it represented a new species of gorilla, but today it is considered a subspecies of gorilla.

The last of the known apes to be described was the pygmy chimpanzee, *Pan paniscus*, discovered by Schwarz in 1929. He considered it to be a subspecies of the common chimpanzee (more on this in the next chapter.) The pygmy chimpanzee lives in the rain forests of central and western Zaire and today is known as "bonobo." Reynolds (1967), however, as mentioned earlier, believes the pygmy chimpanzee may actually be the animal described by Nicolaas Tulp in 1641.

During the 1920s and 1930s primates were beginning to be thought of as objects not only for comparative anatomical studies but also for physiological and psychological investigations. It was being accepted more and more that monkeys and apes were our closest relatives and could therefore be useful in medical research. It was also realized that primate breeding facilities were necessary in order to have a continuous supply of healthy animals for the various studies and experiments that were just on the horizon.

One of the earliest primate stations and breeding colonies was established in Russia (with four monkeys) on August 24, 1927. This is now the world-famous Sukhumi Monkey Breeding Station and has the oldest baboon breeding colony (Lapin and Fridman 1975). Shortly after that, Robert M. Yerkes, a professor of psychology at Yale University, opened the Orange Park Breeding Station in Florida, in July 1930, and in September 1930 its first chimpanzee was born. This primate center has been in continuous operation ever since. It was moved to Emory University in Atlanta, Georgia, in 1965 and now is known as the Yerkes Regional Primate Center.

One other important breeding colony was established in the late 1930s on Cayo Santiago Island off the coast of Puerto Rico. In December 1938, 409 rhesus monkeys were brought from India and put on the island by the psychologist C. R. Carpenter. The colony has been maintained and is still an active rhesus monkey breeding colony and research facility.

The use of primates in medical research has led, among other things, to the discovery of the Salk polio vaccine; medical, behavioral, and other investigations are still providing valuable information.

The European discovery of monkeys and apes began some four thousand years ago. Most of the world has now been explored, and worldwide communications are virtually complete. It is doubtful that any new species of apes will be discovered unless someone uncovers indisputable evidence of the mystical Abominable Snowman, or Yeti, of the high Himalayas, or its counterpart in western North America, the Sasquatch, or Big Foot. It would not be scientific to be complacent, however, and as new species of primates have only recently been found in the jungles of Brazil and lemurs in Madagascar, there may be some surprises ahead for us after all.

2 ∽

Classification

and

Distribution

of Living

Primates

PRINCIPLES OF TAXONOMY

Perhaps the most basic tenet of science is that nature itself is orderly. An understanding of this order is at the core of the physical sciences, and it is no less important for the natural sciences. Humankind has sought to organize and classify the natural flora (plants) and fauna (animals) for millennia. The recognition of the importance of ordering the various components of nature certainly goes back to Aristotle and, as we have seen in chapter one, under Linnaeus's critical eye it became a permanent part of all biological studies. There are three terms that are usually used for this branch of science: systematics, taxonomy, and classification.

These terms are often used interchangeably, although systematics is usually considered to include taxonomy as well as classification. Thus, systematics is "the scientific study of the kinds and diversity of organisms and all relationships among them," and taxonomy is "the theoretical study of classification, including its bases, procedures and rules" (Simpson 1962, pp. 7 and 11, and Martin 1994). Since we are interested in the classification of primates—that is, the ordering of primates into groups based on their relationships—we will use the term "taxonomy" to describe the procedures used in this chapter.

Taxonomy is important for several reasons. In the first place, it is a method by which scientists can organize and order data for better understanding and communication. Second, since Darwin's *Origin of Species* in 1859, it is believed by most scientists that a classification should reflect, as much as possible, biological and evolutionary relationships, that is, that all living creatures are derived from common ancestors who have changed through time by on-going evolutionary processes. This approach is referred to as evolutionary systematics and is based on the phylogenetic relationships among the organisms involved. It is this method of classification that is used in this book (see below for other approaches to classification).

The basis of classification, as noted above, is the degree to which animal groups are related to each other through descent from a common ancestor. Thus, the major problem for taxonomists is to distinguish between those similarities resulting from inheritance from a common ancestor and those that are not inherited from a common ancestor. This is determined by structural similarities based on common descent, homologies. The bones of the forelimbs in the three vertebrates in figure 2.1 are similar in structure and arrangement, and the most logical explanation is that the similarities are due to descent from a common, homologous ancestor. There are also structures that have a similar function but that are not derived from a common ancestor and are known as analogous structures (fig. 2.2).

Another important process that must be identified by taxonomists is parallel evolution. Parallelism is the independent evolution of similar traits in separate lineages. As we will shortly see, certain structural similarities between Old World and New World monkeys are the result of parallel evolution; that is, the two groups or lineages have lived in similar, but separate, tropical arboreal (tree-dwelling) environments since their common ancestry and have evolved similar morphologies so that they are now similar in appearance. Thus, the average person visiting a zoo would probably have difficulty telling them apart. Sorting out the effects of parallel evolution from derived features (new features evolved in a recent common ancestor and retained by its descendants and absent from any ancestral group) is difficult but important when attempting to establish biological relationships among primates.

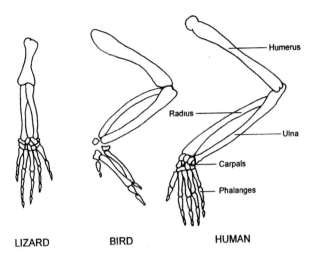

Humerus

Radius

Ulna

Carpals

Phalanges

LIZARD   BIRD   HUMAN

2.1 Homology. The limbs are homologous in that they are composed of comparable bones from a common ancestor. In each animal the limbs are adapted for special functions by differences in length and shape of the individual bones.

Since the publications of Hennig (1966) and Eldredge and Cracraft (1980), the cladistic school of classification has become an important force within systematics and contrasts with the more traditional evolutionary systematics. The advocates of this approach to classification insist that classification should directly reflect the inferred pattern of phylogenetic relationships among members of the groups being classified, that is, that the traits held in common by these members are the result of direct ancestor-descendant relationships. In other words, all members in a group should be monophyletic, indicating that all members are descended from a single stem species. For example, in figure 2.3 (a) C has diverged before A and B, which share a more recent common ancestor; (b) B has separated prior to A and C, which now have the most recent common ancestor; and (c) A has come off first leaving B and C with the most recent common ancestor.

One important difference between cladistics and evolutionary systematics, although both are based on the phylogenetic relationships of the animals involved in the classification, is that evolutionary systematics does not require that it be based exclusively on the inferred branching points (nodes) within the phylogenetic tree. Some groups will be monophyletic, but some will not. It should be mentioned that there is a third way of approaching taxonomy, which is known as numerical taxonomy (Sneath and Sokal 1973). This is based solely on similarities

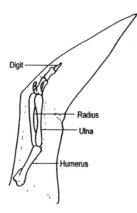

2.2 Analogy. The wings of insects and vertebrates have the same function, but they are not derived from a common ancestor.

Digit

Radius

Ulna

Humerus

Bird

Butterfly

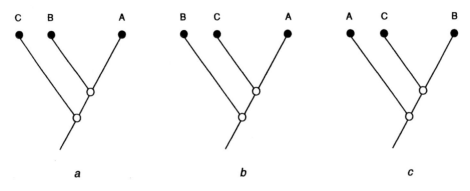

a   b   c

2.3 Three different cladograms showing ancestor-descendant relationships. In each case the evolutionary relationships among the three groups are different.

between species and does not consider that classification need reflect inferred phylogenetic relationships.

The term "taxon" (pl. taxa) is used for a group of organisms at any level of a hierarchic classification. Thus, Mammalia (a class), Primates (an order), Cercopithecidae (a family), *Cercopithecus* (a genus), and *Homo sapiens* (a species) are all taxa. Much of taxonomy is concerned with defining taxa and trying to understand the relationships among them.

Classification and Nomenclature

As discussed in the previous chapter, Carolus Linnaeus developed a system of classification and nomenclature in the eighteenth century that is still in use today. In the tenth edition of *Systema naturae* in 1758, Linnaeus wrote that all plants and animals could be classified by a descending series of categories or hierarchies. The object of any classification is to construct a hierarchical system of categories of increasing rank. Today, the system is known as the Linnaean hierarchy and consists of a basic sequence of seven levels:

Kingdom: Animal
   Phylum: Chordata
      Class: Mammalia
         Order: Primates
            Family: Cercopithecidae
               Genus: *Macaca*
                  Species: *mulatta*

Linnaeus did not use Kingdom and Family; these were added later. It is important to remember that this sequence, from top to bottom, represents a decreasing inclusiveness of each level.

Linnaean nomenclature is often referred to as the binomial or binary system since the genus (pl. genera) is one word and the name of the species (pl. also species) consists of two words. For example, the genus *Macaca* refers to several different species, while the species, *Macaca mulatta* refers only to the rhesus monkey. Generally primates have a common or vernacular name often used after the Latin name, such as *Macaca mulatta*, the rhesus macaque, or *Macaca nemestrina*, the pig-tailed macaque. Additionally, there is what is called a type specimen (a skull, tooth, or skin) that is used to define the species and is the only standard for identifying other specimens as belonging to this same species.

The basic unit of classification, the biological species, is usually defined as a group of actually or potentially interbreeding natural populations that is reproductively isolated from other such groups. This definition is usually satisfactory for living species of vertebrates but is impossible for identifying extinct primate species in the fossil record (see chapter 10). Above the species level, that is, at genus level and above, definitions become quite arbitrary. For example, a genus is defined as a category that groups together closely related species, as opposed to a family, in which all members have certain features in common. Two other levels are occasionally used by taxonomists. These are cohort, positioned between class and order, and tribe, which is usually found between family and genus.

Today it is customary to add levels designated as super above basic levels and sub and infra below them. Also, there are three suffixes which are customarily used to indicate subfamilies: -inae; families: -idae; and superfamilies: -oidea, each added to the stem of the name of one of the genera.

Taxonomic names above the genus are capitalized but not italicized. Thus:

Superfamily: Cercopithecoidea
Family: Cercopithecidae
Subfamily: Cercopithecinae

What Is a Primate?

Primates are mammals that separated from a primitive group of mammals (perhaps the small, primitive insect-eating mammals, the Insectivores) some 60 million years ago. Mammals are one of five taxa (classes) of vertebrate animals. Vertebrates are animals with a vertebral column; this includes not only mammals but fish, amphibians, reptiles, and birds. Mammals are characterized as warm-blooded vertebrates having an advanced reproductive system, mammary (milk) glands, and fur or hair somewhere on their bodies. Birds, incidentally, are the only other warm-blooded vertebrates.

It is difficult to define primates as a group since they lack a single characteristic that separates them from other mammalian orders, as, for example, the gnawing incisor teeth of rodents; the specialized limbs and reduced digits of horses, cows, deers, and pigs; or the wings of bats. Because of this, most authors have looked for evolutionary trends or total morphological patterns when attempting to define and classify primates. These trends and grades of morphological organization of non-human primates have usually been compared with those of human organization. This is satisfactory for certain types of studies, but a more instructive approach for understanding primate evolution might be gained by comparing primates with other animals that have become adapted to similar life styles (Cartmill 1974 and 1994). Whichever approach one uses, it seems clear that primates have remained generalized, retaining many primitive features during their evolutionary history rather than becoming highly specialized. Consequently, they are better defined by their adaptability than by their adaptations. As noted by Gingerich (1984), in order to understand the grouping (classification) of primates, as with other orders, it is necessary to look at their evolutionary history. We will consider this viewpoint more thoroughly in chapter ten.

The first major definition of the primates was published by the British anatomist St. George Mivart in 1873. He wrote that Primates as an order are "unguiculate, claviculate placental mammals, with orbits encircled by bone; three kinds of teeth, at least at one time of their life; brain always with a posterior lobe and a calcarine fissure; the innermost digit of at least one pair of extremities opposable; hallux with a flat nail or none; a well-developed cecum; penis pendulous; testes scrotal; always two pectoral mammae" (p. 490).

It is important to remember that none of these anatomical features is unique to primates, but that, when taken together, they characterize the order. It should be pointed out, however, that this characterization is essentially limited to living primates since it is based to a large extent on the anatomy of soft tissues and has little use in identifying fossil primates. The current definition of primates still uses many of the anatomical characteristics proposed by St. George Mivart, although several additional criteria have been added through the years by various primatologists (notably, Le Gros Clark 1971 and Napier and Napier 1967).

The current definition of primates is as follows: primates have retained a generalized skeleton, with five fingers and toes (pentadactyly). The innermost digit on the hand, the thumb (pollex), and foot, the big toe (hallux), usually has enhanced mobility, and there is a tendency for nails to replace claws on at least some of the digits. There are also touch (tactile) pads possessing finger and toe prints at the ends of the digits. The orbits are encircled with bone, that is, there is a postorbital bar present so that the eye is completely surrounded by bone. A characteristic of all primates is that the middle ear chamber, or auditory bulla, is formed from the petrous portion of the temporal bone. There is retention of the clavicle (collarbone) in all primates, which enhances the mobility of the shoulder joint. The teeth tend to be generalized, especially the cheek teeth, and most living primates have two incisors in each half of each jaw. There has been an increase in visual depth perception with the development of binocular vision as the eyes have moved forward and closer together on the skull. Color vision is probably present to some degree in all primates except some of the nocturnal (active at night) species. A general increase in the size and complexity of the brain has occurred during primate evolution, with the result that the primate brain, relative to body weight, is the largest of all land mammals. There are typically two pectoral breasts, an enhancement of fetal nourishment, and an increase in placental complexity. Males have a pendulous penis, the testes are scrotal, and all primates have a cecum at the beginning of the large intestine. Most primates give birth to a single offspring. And finally, there is a trend toward longer gestation times, as well as an extension of the postnatal life span in the higher primates. Each of these definitions will be more thoroughly discussed in subsequent chapters.

Primate Classification

Primate classification has always been in a state of flux, and probably always will be. There is continual disagreement over many parts of the classification, especially over which genera should be admitted to the order at one end of the hierarchy, and which genera belong in the family Hominidae at the other end of the spectrum. For example, from time to time the small, squirrel-like tree shrew of Southeast Asia has been classified as a primate. The tree shrew, though not considered a primate today, is still important for understanding the anatomical organization of primates because it represents a form probably not far removed from the adaptive level of early placental mammals. This insectivorous creature is placed in its own order, Scandentia, although, along with bats, flying lemurs, and primates, it is part of what is considered to be a single grandorder, Archonta, due to certain basic similarities of morphology among them (McKenna 1975).

Since the early 1960s the conventional classification that put humans in the family Hominidae while putting gorillas, chimpanzees, and orangutans in the family Pongidae has been under attack by biochemical and molecular anthropologists and biologists as well as by cladists. Morris Goodman led the way when he suggested that, based on his analysis of blood proteins, the two African apes belonged in the same family as humans. Within the past several years DNA hybridization studies, which determine how much of one set of DNA fragments will bind to another, thereby indicating how many sequences they share, have pointed to a very close genetic relationship between humans and chimpanzees, with gorillas more distant. Recently, Goodman and his colleagues have sequenced long series of DNA and come to the conclusion that humans and chimpanzees have only a 1.6 percent difference in the noncoding portions of their globin gene clusters, whereas there is a 2.1 percent difference between chimpanzees and gorillas (Gibbons 1990). To some, this implies that the living Hominidae are separated into two subfamilies—the Ponginae, containing only the orangutan, and the Homininae, with gorillas, chimpanzees, and humans (see table 3.6). Again, molecular anthropologists are finding that appearances may be deceiving. Based on recent DNA sequence studies by Maryellen Ruvolo of Harvard University, there is a genetic gap between the western lowland gorilla and the mountain gorilla sufficient to suggest two species (Morell 1994). This particular issue is not closed, however, and as work is continuing in several molecular biological laboratories new results may overturn the present classification at any time.

Several years ago the eminent British anatomist J. Z. Young (1971, p. 416) made an interesting observation regarding the individuals who make primate classifications. I quote it in full: "The sensitivity and disagreement among biologists and anthropologists about such questions provides an amusing insight into their humanity, though it leads to much confusion." This does not make classifications any less important, but simply means that as new methods become available, the findings of biochemical research (e.g., immunologic blood proteins, mitochondrial DNA sequences, and DNA hybridization studies) offer additional new data to the taxonomist for the ordering of primates within the classification as mentioned above. This is why at any given point in time there may be several alternative classifications available to the primatologist. For example, in the classification presented in table 2.1, there are the traditional two suborders, Prosimii and Anthropoidea. Prosimians (lemurs, lorises, and tarsiers) are the "before-apes," or what German primatologists generally refer to as Halbaffen or "half-apes." The prosimians are also often referred to as the lower primates since they have retained a more conservative or primitive morphology and behavior, a condition probably not too unlike that of the primates of the Eocene (see chap. 10). Anthropoids (monkeys, apes, and man), on the other hand, are the "manlike" primates and are frequently termed the higher primates. Recently, however, some primatologists have resurrected an

Table 2.1   Classification of Living Primates

Order: Primates

Suborder: Prosimii

Infraorder: Lemuriformes

Superfamily: Lemuroidea

Family: Lemuridae
   Genera
     Lemur [lemurs]
     Varecia [varigated lemur]
     Hapalemur [gentle lemur]

Family: Lepilemuridae
   Genera
     Lepilemur [sportive lemur]

Family: Cheirogaleidae
   Genera
     Microcebus [mouse lemur]
     Mirza [Coquerel's mouse lemur]
     Cheirogaleus [dwarf lemur]
     Phaner [forked lemur]
     Allocebus [hairy-eared dwarf lemur]

Family: Indriidae
   Genera
     Indri [indri]
     Propithecus [Verreaux's sifaka]
     Avahi [avahi or woolly lemur]

Family: Daubentoniidae
   Genus
     Daubentonia [aye-aye]

Superfamily: Lorisoidea

Family: Lorisidae
   Genera
     Loris [slender loris]
     Nycticebus [slow loris]
     Arctocebus [golden potto]
     Perodicticus [potto]

Family: Galagidae
   Genera
     Otolemur [large-eared bush baby]
     Galago [bush baby]
     Galagoides [dwarf bush baby]
     Euoticus [needle-clawed bush baby]

Infraorder: Tarsiiformes
Family: Tarsiidae
   Genus
     Tarsius [tarsier]

Suborder: Anthropoidea

Infraorder: Platyrrhini

Superfamily: Ceboidea

Family: Cebidae

Subfamily: Cebinae
   Genera
     Cebus [capuchin monkey]
     Saimiri [squirrel monkey]

Subfamily: Aotinae
   Genera
     Aotus [night monkey]
     Callicebus [titi monkeys]

Subfamily: Atelinae
   Genera
     Ateles [spider monkeys]
     Lagothrix [woolly monkeys]
     Alouatta [howler monkeys]
     Brachyteles [woolly spider monkey]

Subfamily: Pitheciinae
   Genera
     Cacajo [uakaris]
     Chiropotes [bearded sakis]
     Pithecia [sakis]

Family: Callitrichidae

Table 2.1   Classification of Living Primates (continued)

Subfamily: Callitrichinae
   Genera
      Saguinus [tamarins]
      Leontopithecus [lion tamarin]
      Callithrix [marmosets]
      Cebula [pygmy marmoset]
      Callimico [Goeldi's marmoset]

Infraorder: Catarrhini

Superfamily: Cercopithecoidea

Subfamily: Cercopithecinae
   Genera
      Macaca [macaques]
      Cercocebus [mangabeys]
      Papio [baboons]
      Mandrillus [mandrills]
      Theropithecus [gelada baboon]
      Cercopithecus [guenons and vervets]
      Erythrocebus [patas monkey]
      Miopithecus [talapoin]
      Allenopithecus [Allen's swamp
      monkey]

Subfamily: Colobinae
   Genera

Colobus [guerezas]
Presbytis [langurs]
Pygathrix [douc langurs]
Simias [Pagai Island langur]
Nasalis [proboscis monkey]
Rhinopithecus [snubnosed langurs]

Superfamily: Hominoidea

Family: Hylobatidae
   Genera
      Hylobates [gibbons and siamang]

Family: Pongidae

Subfamily: Ponginae
   Genera
      Pongo [orangutan]

Subfamily: Gorillinae
   Genera
      Gorilla [gorilla]
      Pan [chimpanzees]
Family: Hominidae
   Genera
      Homo [humans]

alternative scheme proposed by Pocock (1918) that places the tarsiers with the anthropoids in the suborder Haplorhini, while placing lemurs, lorises, and galagos in the suborder Strepsirhini. The basis of this classification is the anatomy of the nose. The haplorhines possess nostrils surrounded by hairy, dry skin, whereas the nostrils in strepsirhines are surrounded by moist skin without hair, the rhinarium, which has a philtrum that attaches the upper lip to the gums (fig. 2.4). This book treats tarsiers as prosimians while recognizing the mosaic nature of their evolutionary development. Mosaicism refers to the fact that different structural systems in an organism may evolve at different rates through time resulting in differential evolutionary development, as exemplified by living tarsiers.

Many of the traits possessed by primates are generally considered to be the result of life in the trees. The hands and feet are good examples

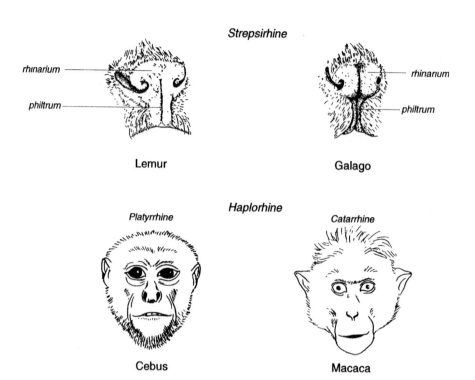

2.4 Basic differences in the structure of the primate nose. The nose of strepsirhine primates (all prosimians except *Tarsius*) is composed of a moist rhinarium, which, as part of the upper lip, is attached to the gum by the philtrum. The rhinarium is a complex structure serving both smell and touch (vibrassae or "whiskers"). The nose of haplorhine primates (*Tarsius* and all Anthropoidea) lacks the moist rhinarium. The upper lip is free and mobile and contains an important muscle of facial expression. (Adapted from Hershkovitz 1977)

The nose of all New World monkeys (platyrrhine) is broad and the nostrils are widely separated and directed laterally, while in Old World monkeys, apes, and humans (catarrhine) the nostrils are close together and point downward.

of these adaptations (fig. 2.5 A, B, C, and D). The thumb in most primates is separated from the other digits and displays some degree of opposability, as in our own hands. The big toe is opposed to the other digits of the foot by a cleft in all nonhuman primates, which results in an excellent grasping organ for use in trees. The hallux of the human foot is the only exception among living primates, but even here the muscles are the same as those in the nonhuman primate foot. The close-set eyes with binocular vision, along with the reduction of smell,

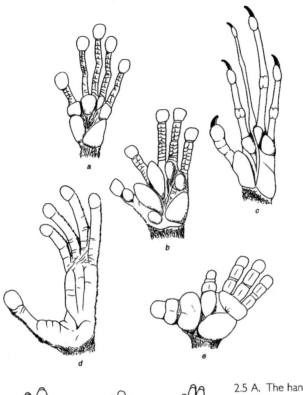

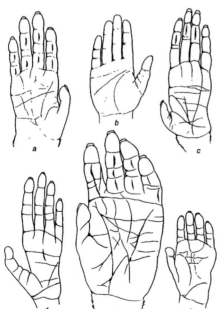

2.5 A. The hands of prosimians:
a) *Tarsius*; b) *Galago*;
c) *Daubentonia*; d) *Indria*;
e) *Nycticebus*.

2.5 B The hands of anthropoids.
a) *Pan*; b) *Homo*; c) *Pongo*;
d) *Hylobates*; e) *Gorilla*; f) *Papio*.

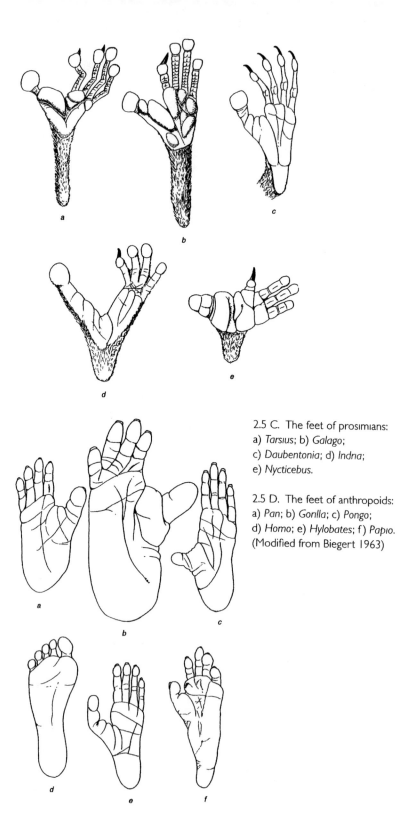

2.5 C. The feet of prosimians:
a) *Tarsius*; b) *Galago*;
c) *Daubentonia*; d) *Indria*;
e) *Nycticebus*.

2.5 D. The feet of anthropoids:
a) *Pan*; b) *Gorilla*; c) *Pongo*;
d) *Homo*; e) *Hylobates*; f) *Papio*.
(Modified from Biegert 1963)

also suggest tree-dwelling. This complete suite of tree-dwelling primate adaptations has for years formed the basis of what is generally known as the arboreal theory of primate origins. But has this theory been accurate?

Matt Cartmill (1972, 1974, and 1994) of Duke University has suggested that the close-set eyes and grasping hands with nails, as well as other primate characteristics, may not have begun as arboreal adaptations. Rather, they may represent adaptations for a visual-predation mode of life in trees and bushes by small ancestral primates searching for insects. This visual-predation theory of primate evolution has also been dubbed the "bug snatching" hypothesis, and some scientists believe the two theories are not mutually exclusive (Nelson and Jurmain 1991). Indeed, Nelson and Jurmain see visual predation as a possible preadaptation for arboreal life—that is, when the primates became arboreal, the visual-predation adaptations were already well suited for this mode of life, and they simply became part of their arboreal adaptations. As we shall see in chapter ten, there is fossil evidence to suggest that primates were already arboreal by the Eocene, some 50 million years ago.

The living primates may be grouped into six natural groups consisting of 1) the lemurs of Madagascar; 2) the lorises and bush babies of Africa and Asia; 3) the tarsiers of Southeast Asia; 4) the New World monkeys of Central and South America; 5) the Old World monkeys of Africa and Asia; and 6) the apes of Africa and Asia, and humans (Martin 1994 and fig. 2.6). The primate classification presented here is often referred to as the traditional one and is based on the concept of grade or level of organization of the various animal groups. The term grade is used to indicate animals at similar levels of organization and in no way implies any superiority or inferiority. Also, as the late James Gavan (1975, p. 1) pointed out, primate classifications represent only "A Classification . . . , not The Classification." This distinction is important since it is unlikely that there will ever be a final classification of primates (or of any organisms, for that matter) with which all authorities will agree. The diverse disciplines and the different methods of data collection with the subsequent analyses and interpretations of the data represent one of the major strengths of science: ideas, hypotheses, and classifications are always under scrutiny, with the result that they are always changing or being modified as new information becomes available.

GROUP 1: LEMURS (Madagascar)

Mouse Lemur Subgroup   True Lemur Subgroup   Indri Subgroup   Aye-aye

GROUP 2: LORIS GROUP (Africa and Asia)

Lorises   Bushbabies

GROUP 3: TARSIERS (Southeast Asia)

GROUP 4: NEW WORLD MONKEYS (South and Central America)

"True" New World Monkeys   Goeldi's monkey   Marmosets and Tamarins

GROUP 5: OLD WORLD MONKEYS (Africa and Asia)

Leaf-monkeys   Baboons, Macaques and Guenons

GROUP 6: APES and HUMANS (Africa and Asia)

Lesser Apes   Great Apes   Humans

2.6 Six groups of living primates. Each group has specific geographical distributions and forms the basis of all primate classifications. (After Martin 1994).

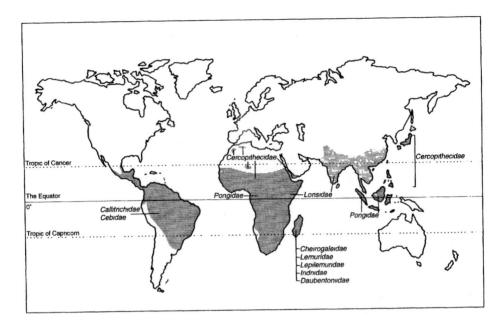

The map labels read:

Tropic of Cancer

Cercopithecidae

Cercopithecidae

The Equator

0°

Pongidae

Lonsidae

Callitrichidae
Cebidae

Pongidae

Tropic of Capricorn

Cheirogaleidae
Lemuridae
Lepilemuridae
Indriidae
Daubentoniidae

2.7 The geographic distribution of extant nonhuman primates.

DISTRIBUTION OF LIVING PRIMATES

The study of the distribution of living animals is zoogeography, and to-day there are about 190 primate species grouped into some 50 genera, most of which inhabit tropical or subtropical regions of the world (fig. 2.7). The majority, some 130 species, are monkeys, of which about 50 species live in the New World and 80 in the Old World. Prosimians make up about 40 species, while there are 14 species of apes.

LEMURS   Madagascar is one of the largest islands in the world, strad-dling the Tropic of Capricorn off the southeast coast of Africa. It is the home today of all living lemurs, and includes the prosimian families Cheirogaleidae, Lemuridae, Lepilemuridae, Indriidae, and Daubentonii-dae. Three other prosimian families—Lorisidae, Galagidae, and Tarsii-dae—live in Africa and Asia. No monkeys or apes ever got across the Mozambique Channel from Africa.

The prosimians of Madagascar represent a diversified group of pri-mates that over the last 50 million years or so have occupied the vari-ous ecological niches of the island. Several species of lemurs are shown in figure 2.8. The Malagasy lemurs, with the exception of the aye-aye,

2.8 Several species of Malagasy Prosimians: a) *Indria*; b) *Microcebus*; c) *Propithecus*; d) *Daubentonia*; e) *Varecia*; f) *Lemur*. (Redrawn from Napier and Napier 1985)

have a dental comb formed by the lower incisor teeth and the canines (see fig. 5.3). They also have a grooming claw on the second digit of the foot (see chap. 9). They range in size from one of the smallest primates, the mouse lemur, *Microcebus* (about 70 to 80 g) to the largest living lemur, *Indri* (about 10.5 kg), which, with their short, stumpy tails, are frequently referred to as the "apes of the prosimian world" (Kavanagh 1984, p. 51). The mouse lemur eats about anything, for example, fruits, flowers, insects, gum, tree frogs, eggs, and small birds. Indris, on the other hand, are the most folivorous (leaf-eating) of the prosimians. Many Madagascar lemurs are classified as vertical clingers and leapers (see chap. 7), for they use their long legs to leap between trees (fig. 2.9) and, when on the ground, may hop kangaroo-fashion from tree to tree or walk quadrupedally (fig. 2.10).

Prosimians may be nocturnal (active at night), diurnal (active during the day), or crepuscular (active at dawn and twilight). Most lemurs are nocturnal, although some of the larger ones, such as the indri, the ring-tailed lemur, and the sifakas are diurnal (fig. 2.11). Tattersall (1979) described a different activity pattern for the Mayotte lemur. These animals are active at various times of the night as well as during daylight hours, inspiring him to propose the term cathemeral (Gr. kata, through, and hemera, day) to describe this activity pattern. The term has since been used by other primatologists to describe this activity pattern in other primates, and so we have added another term to our vocabulary.

The aye-aye is rare, bizarre, and nocturnal (see fig. 2.8 d). For several years after its discovery in 1780, the aye-aye was considered to be a squirrel, probably because it has gnawing, rodentlike incisors. Aye-ayes eat mainly insects, grubs, and larva, which they find by listening or smelling. When they find them, they dig them out of their burrows in tree trunks using their rodentlike incisors and long, slender middle fingers. Since there are no woodpeckers on Madagascar it has been suggested that the aye-aye fills this ecological niche (Cartmill 1974).

LORISES AND POTTOS   The other prosimian taxa occupy regions of Africa and Southeast Asia and belong to three families: the Lorisidae, Galagidae, and Tarsiidae. Like lemurs, the lorisids have a dental comb and a grooming claw on the second digit of the foot. Their diet consists mainly of insects, snails, eggs, lizards, and fruit. The pottos belong to two genera that have slow, deliberate, quadrupedal movements which permit them to creep along branches in their search for food in their nocturnal world in western and central Africa. Arctocebus, the golden potto, is smaller and is also known as angwantibo. This small primate is an interesting combination of characteristics, featuring a foxy face; no tail; small, naked ears; and small, short fingers and toes. Perodicticus, or Potto, is the best known of the pottos and lives in the continuous

2.9 *Lepilemur mustelinus*, a vertical clinger and leaper from Madagascar. (Photo courtesy of Robert W. Sussman)

canopy forests of western Africa. Its range, however, extends from western Africa through central Africa as far east as Kenya.

The slow loris, *Nycticebus*, lives in the jungles of Southeast Asia and some of the western islands of the Indonesian archipelago (fig. 2.12). Its name pretty well describes its locomotion, which is ultracautious when contrasted with that of galagos, as it forages through the forests at night looking for fruit and insects. Another variety of loris, the slender loris, whose genus name is also *Loris*, is a nocturnal creeper inhabiting southern India and Sri Lanka. The slow loris appears to be mainly a fruit eater (frugivorous), while the slender loris seems to prefer insects (insectivorous).

GALAGOS The galagos, or bush babies, have an extensive geographic distribution throughout much of sub-Saharan Africa (fig. 2.13). Galagos are represented by three or four nocturnal genera, and all are excellent leapers between trees, as well as on the ground. Accordingly,

2.10 When on the ground *Lemur catta* from Madagascar may walk quadrupedally as seen here. (Photo courtesy of Robert W. Sussman)

2.11 *Propithecus verreauxi*, clinging to a tree in Madagascar. (Photo courtesy of Robert W. Sussman)

2.12 *Nycticebus coucang*, one of two nocturnal species of slow loris living in Southeast Asia. Notice the short second digit on the hand and foot. (Photo courtesy of Lee Haines, Woodland Park Zoo, Seattle)

2.13 *Galago senegalensis*, one of several species of bush babies living over much of Africa south of the Sahara. Notice the folds on the inner side of the ears. (Photo courtesy of Lee Haines, Woodland Park Zoo, Seattle)

their mode of locomotion is quite different from that of the pottos. Bush babies eat a variety of foods, and depending on the species their diet includes insects, frogs, snails, fruit, and gums that they scrape off trees. The western African needle-clawed galago, *Euoticus elegantulus*, or "Elegant Ones with Nice Ears," has a name that apparently belies its temperament

2.14 *Tarsius bancanus*, one of three species of tarsiers living on many of the islands off Southeast Asia. Notice the large eyes, dry rhinarium, short trunk, long legs, and very long tail. The animal is poised in a vertical clinging position. (Photo courtesy of Nina G. Jablonski)

which, according to Sanderson (1957, p. 21), is "aggressive, most furious and irascible."

TARSIERS   The tarsiers, here considered as a family (Tarsiidae) within the Prosimii, are also found in Southeast Asia and the Philippine Islands (fig. 2.14). These small primates are supreme leapers. Jumping up to fifteen feet, they land vertically on the side of a tree and immediately look for prey. There is only one genus, *Tarsius*, with perhaps four or five species. They are nocturnal primates living completely on small animals (faunivorous) such as frogs, lizards, snakes, and insects.

Prosimians were much more widely distributed during the early Cenozoic era but became more circumscribed through time as the climate changed (a general lowering of the mean temperature), as competition increased after the appearance and spread of the higher primates, and, finally, as humans appeared on the scene. There is little doubt that humans have played a major role in the geographical retreat of the prosimians on Madagascar, as well as in other regions of the world (see chap. 11).

New World Monkeys

The New World monkeys (platyrrhines or broad-nosed monkeys) are all included in the superfamily Ceboidea (table 2.1), but, as we will see later, there is considerable discussion concerning their classification at the family and specific levels. The New World monkeys, like all anthropoids, represent a single clade (all members descended from a common ancestor): in this case, a single anthropoid ancestor. It is customary to separate them into two families: Callitrichidae (marmosets and tamarins) and Cebidae (all other New World monkeys). Some primatologists would have a third family, the Callimiconidae, which includes one genus, *Callimico goeldii*, Goeldi's monkey, while others have put it with the Cebidae. Goeldi's monkey has characteristics that suggest membership in both Callitrichidae and Cebidae, which helps to explain the difficulty surrounding its classification.

The New World monkeys represent a diverse group of tropical, arboreal primates living today in the rain forests of Central America and Mexico, and through much of South America as far south as Argentina (fig. 2.7). They range from along the east coast of South America to the Peruvian Andes and from sea level to approximately 3,200 meters (10,500 ft) in altitude. They are all diurnal with one exception: *Aotus* is the world's only nocturnal monkey.

MARMOSETS AND TAMARINS   The marmosets and tamarins (Callitrichids) are the smallest of South American monkeys and have claws on all their digits except the big toe, which has a broad, flat nail (fig. 2.15). With the exception of *Callimico*, they also lack their upper and lower third molars, the so-called wisdom teeth of humans. All species are diurnal, are quadrupedal when running through the trees, and show little difference in body size between males and females. Their diets consist mainly of insects and fruit. Twins are common, and births may occur twice a year. Both sexes take care of the young, and while traveling, the males normally carry the young.

The only primates with prehensile tails are found among the cebids. Prehensile tails have tailprints on their under surfaces similar to those present on the hands (fingerprints), and the tails are used for swinging

2.15 The silvery marmoset, *Callithrix argentata*, lives in Amazonia. Marmosets are the smallest New World monkeys. (Redrawn from Few 1991)

from branch to branch as well as for holding objects (fig. 2.16). In fact the only platyrrhine without a long tail is the *Cacajao*, or uakari monkey, which has a rather short, bushy tail. Platyrrhines have remained arboreal throughout their various habitats and have come to occupy most of the rain forests of South America. There are no terrestrial South American monkeys, but some species do spend short periods on the ground, for example, the black howler monkey, *Alouatta caraya*.

CAPUCHINS   The capuchin monkey (*Cebus*) is widely spread throughout forests from Honduras to Argentina. They eat a variety of foods including insects, leaves, fruits, and on occasion small birds. They play an important role in forest ecology by pollinating flowers as they move through the forests, but in addition they also help to disperse seeds from the fruits they eat and pass through their digestive systems. Incidentally,

2.16 The spider monkey, *Ateles*, swinging forward from a tree branch using its prehensile tail, which has a large friction pad (tailprint) on the underside of the distal portion. The thumb is reduced or absent as shown here. (Redrawn from Napier and Napier 1985)

according to Kavanagh (1984), many seeds stand a better chance to germinate when they pass through the digestive system of monkeys since the seeds have been kept warm and moist and exposed to nutrients. Thus, forests benefit from the presence of monkeys. Capuchins are considered to be quite intelligent and adapt quickly to new surroundings, and perhaps because of this versatility they are the "organ-grinder" monkeys we often see in old movies.

SAKIS AND UAKARIS   The pithecine monkeys are the sakis (*Pithecia*) and uakaris (*Cacajao*) (fig. 2.17). The sakis are well known for their locomotor skills and in some regions are known as flying monkeys. They rarely come to the ground, preferring the middle levels of the forest. Their diet is mainly fruit and seeds. The shaggy uakaris have short, bushy tails. Their diet consists of fruit, leaves, and flowers. They are rare, living in limited areas in the rain forests of the Amazon basin. One species of the uakaris monkey is baldheaded, with red hair and a red face.

2.17 *Pithecia pithecia*, the saki monkey, lives in many of the rain forests of South America. There is a great deal of sexual dichromatism (having two color varieties) between the sexes, as shown here. The male is on the left and the female on the right. (Photo courtesy of Robert W. Sussman)

SPIDER AND WOOLLY MONKEYS The spider monkey (*Ateles*) and the woolly monkey (*Lagothrix*) have long, prehensile tails, and their thumbs are greatly reduced or absent, particularly in the spider monkey. They travel by running along tree branches or by swinging below them using their arms or tails. The woolly monkey eats leaves, fruit, and flowers. The spider monkey is very agile in the trees. With its long limbs and extremely dextrous prehensile tail, it is capable of moving quickly through the trees (see fig. 2.16). Its hands are really no more than hooks since the fingers are elongated and the thumb is either absent or much reduced. Its diet is mainly fruit.

HOWLER MONKEYS Howler monkeys (*Alouatta*) are among the largest New World monkeys, with males weighing between 7 and 9 kilograms (15–20 lb). Their tails are prehensile, and they often use them to hang by while they reach out to the more succulent leaves on the terminal ends of tree branches. Of all New World monkeys, the howler is the most folivorous, although fruits are included in its diets from time

2.18 Red howler monkeys, *Alouatta seniculus*, howling at a neighboring troop in the forests of Venezuela. Adult female (left), adult male (center), and large juvenile (right). (Photo courtesy of Carolyn M. Crockett)

to time. As with all New World monkeys, it rarely comes to the ground (fig. 2.18).

SQUIRREL MONKEYS The common squirrel monkey (*Saimiri sciureus*) is found in many of the tropical forests of South America. It is the smallest of the cebids, weighing between a pound and a pound and a half for both sexes. Its small, round face is white with a dark patch around the mouth (fig. 2.19). Its diet consists mainly of fruits and some insects, and like all platyrrhines except the night monkey, it forages during the day.

NIGHT MONKEYS The night monkey (*Aotus*) is also called the owl monkey or douracouli, and, as noted above, it is the only nocturnal anthropoid primate. Night monkeys are rather small (about 1 kg or 2 lb), and there is little size difference between the sexes. They live in tropical forests from Panama to Argentina, subsisting mainly on fruits, but also eating insects and leaves. They are monogamous, living in groups of two to four individuals. Night monkeys are closely related to the small titi monkey (*Callicebus*), and the two genera are placed in the subfamily Aotinae.

2.19 *Saimiri sciureus*, the squirrel monkey of South America, is one of the smallest Cebidae. As shown here, the muzzle is always dark. (Photo courtesy of Robert W. Sussman)

Old World Monkeys

The Old World monkeys are catarrhines (narrow-nosed), as are apes and humans (see fig. 2.4). Old World monkeys have spread throughout Africa and Asia in one family, Cercopithecidae, and two subfamilies, Colobinae (leaf-eating) and Cercopithecinae (non-leaf-eating), each of which is found in Africa and Asia (fig. 2.7). Old World monkeys have adapted into many different ecological niches and are not limited to the tropical and subtropical forests, as are the platyrrhines; instead, several African species have moved into open country and the savannahs, while others have spread into the montane grasslands. As we will see, many of the anatomical differences between the two subfamilies are related to their diet and behavior.

COLOBUS MONKEYS   There is one genus of leaf-eaters in Africa, *Colobus*, represented by several species that live in the forests of western Africa, central Africa, and separate regions of eastern Africa (fig. 2.20). Colobine monkeys are all leaf-eaters and have evolved various methods of extracting the maximum amount of nutrients from the leaves (some

2.20 The black colobus, *Colobus satanas*, lives in western and central Africa and is less folivorous than other colobus monkeys. (Photo courtesy of Robert W. Sussman)

of these are discussed later in chap. 5). Two anatomical features should be noted: all colobines have high-crowned molars with sharp cusps for slicing and cutting leaves and have a sacculated (chambered) stomach for the slow passage of large quantities of fermenting leaves through the gut (see chap. 5, fig. 5.11).

There are several taxa of African colobus monkeys. The largest is the black and white colobus, or guereza. The males weigh between 9 and 15 kilograms (20–32 lb), while the females range between 7 and 10 kilograms (15–22 lb). They occupy the main canopy levels of the forest and subsist mainly on leaves, although fruits are important additions to their diets. There are also the black and the red colobus monkeys as well as the little-known olive colobus monkey from western African forests. The olive colobus monkey is the smallest of the African colobus monkeys, the male weighing between 3 and 4 kilograms (6.5–8.8 lb), and the female slightly less. They live in the lower, more dense levels of the forests. All colobus monkeys have one anatomical feature in common: the thumb is absent or greatly reduced. In fact, the name colobus comes from the Greek word *colob*, which means shortened or mutilated. The result is a hooklike hand, similar to that of the New World spider monkey (see fig. 8.7), that is well adapted for swinging quickly from branch to branch.

All other leaf-eating monkeys live in India, China, Malaysia, Indonesia, Borneo, Sumatra, and Java. These include such diverse species as the hanuman langurs (*Presbytis*), the proboscis monkey (*Nasalis*), and the snub-nosed monkey (*Rhinopithecus*).

LANGURS   The hanuman langurs are the sacred monkeys of the Hindu religion (fig. 2.21). They are protected in India and through the years have come to live near or in the villages and smaller towns. They are the largest and most dimorphic of the langurs, with males weighing about 18 kilograms (40 lb) and females about half that at 11 kilograms (25 lb). In contrast to the colobus monkeys of Africa, langurs have thumbs, but short ones. Although arboreal, they spend much of their day foraging on the ground and are the most terrestrial of all colobines (fig. 2.22). In general their diet consists of more fruits and flowers than does that of most other langurs, although they still eat leaves, preferring the young, tender ones to the more mature leaves. The social life of the hanuman langurs has become well known since the studies of Jay (1965) and Hrdy (1977).

There are at least 15 other langur species living in the forests of India, Southeast Asia, and Indonesia. These forest langurs are somewhat smaller than the hanuman langurs, and there is also less sexual dimorphism. They are all arboreal and folivorous. In some regions two species live together and share the same forest since they have different feeding and activity strategies.

PROBOSCIS MONKEYS   The proboscis monkey of Borneo (*Nasalis larvatus*) is distinguished by its long, pendulous nose, which is particularly well developed in males (fig. 2.23). Its function is still unknown, although it has been suggested that the females prefer males with long noses, and if this is so, this would be a good example of sexual selection. Males can weigh 24 kilograms (52 lb) and are about twice as large as females at 12 kilograms (26 lb), which makes this species among the most dimorphic of all monkeys. They live along rivers in swamp forests and are very good swimmers. They have been observed to dive (jump) into the water from as high as 50 feet (Kavanagh 1984).

2.21 *Presbytis entellus* is the sacred or hanuman langur of India and Sri Lanka. (Photo courtesy of Robert W. Sussman)

2.22 Hanuman monkeys relaxing in a forest in India. The hanuman langurs are considered to be the most terrestrial of all colobine monkeys. (Photo courtesy of Robert W. Sussman)

2.23 The well-known proboscis monkey of Borneo, *Nasalis larvatus*. As shown here, the males are much larger than the females and have large pendulous noses compared to the smaller proboscis of the females. (Redrawn from Few 1991)

SNUB-NOSED MONKEYS   The snub-nosed monkey (*Rhinopithecus roxellana*) is also known as the golden monkey (fig. 2.24 a and b). The monkey's specific name, *roxellana*, commemorates Roxellana, the favorite wife of Suleiman the Magnificent, who was known as the "laughing one." The golden monkey frequently appears to be smiling (Simon and Geroudet 1970). In addition to the snub nose, adult males have fleshy flaps of skin hanging from the sides of their upper lips with no known function. Living in the high mountains of various remote regions of China, this animal is still relatively unknown, although this is beginning to change (Jablonski and Peng 1993). They are large monkeys, the males weighing 15 kilograms (33 lb) or more, the females about 10 kilograms (22 lb). Their diet consists of young bamboo shoots, as well as

2.24 a and b. These pictures were taken of *Rhinopithecus roxellana*, the golden or snub-nosed monkey, near Deqin Village in northwestern Yunnan Province, China. (Photos courtesy of Zhong Tai/Xiao Lin)

leaves, fruits, and buds. It is a highly endangered species, and one reason for this is that its fur was, and perhaps still is, considered a preventive against rheumatism (Simon and Geroudet 1970). Today, however, it is protected by the Chinese government.

DOUC MONKEYS   The douc monkey (Pygathrix) of Southeast Asia is little known. It is one of the more colorful monkeys, with a yellowish face surrounded by a white beard and a body covered with grayish hair. Its thighs are dark to the knees, where the hair becomes more maroon over the legs. The hair of the wrists and hands is white, as though the animal is wearing white gloves. The tail is long and white. The douc monkey is about the size of a langur, and the males are somewhat larger than the females (Kavanagh 1984). They subsist mainly on leaves, fruit, and buds.

The cercopithecines also live throughout much of Africa and Asia. They have colonized various environments, including rain forests, arid scrublands, grasslands, and the high slopes of mountains. Several species are terrestrial, particularly baboons in Africa and macaques in Africa and Asia (fig. 2.25). They eat about anything that is edible in their environments, and some species are truly omnivorous (feeding on both animal and vegetable substances). All cercopithecines differ anatomically from the colobines in two important features: one, their stomachs are simple, lacking the sacculations present in the colobines; and two, they have cheek pouches for temporary food storage (see chap. 5).

MACAQUES   The genus Macaca has the widest geographical distribution today of any Old World monkey, from Morocco in northwestern Africa to the northern end of Honshu Island in Japan at almost 41 degrees north (see fig. 2.7). At this latitude it has the dubious honor of being the only nonhuman primate living that far north and is frequently referred to as the Japanese snow monkey, Macaca fuscata (fig. 2.26). During the winter months it grows a thick covering of hair and subsists mainly on the bark of trees and the occasional invertebrate it finds living in the narrow microenvironment between the snow and the surface of the ground.

Macaques are medium sized, males are always larger than females, and all species are both arboreal and terrestrial (fig. 2.27). Their diet is mainly fruit, leaves, and small invertebrates. Perhaps the best-known macaque is M. mulatta, the rhesus macaque, because of its use as a laboratory animal in much of the Western world. India prohibited their

2.25 Baboons, *Papio cynocephalus*, sunning themselves on a rock. Male in front and female in rear. (Photo courtesy of Judy Johnson, Regional Primate Research Center, Seattle)

2.26 *Macaca fuscata*, or Japanese macaque, ranges farther north than any other nonhuman primate species. At these latitudes, about 40 degrees N, it is frequently referred to as the Japanese snow monkey. (Photo courtesy of Ellen Ingmanson)

2.27 *Macaca fascicularis*, the crab-eating macaque of Southeast Asia, eats not only crabs but a variety of other invertebrates as well as many small vertebrates. (Photo courtesy of Robert W. Sussman)

export in 1977, and many laboratories in the U.S. have now established breeding colonies (see chap. 11). Incidentally, the only Old World monkeys lacking tails are found among the macaques. They are M. *sylvanus*, the Barbary macaque, and M. *nigra*, the Celebes black macaque. Interestingly, the Barbary macaque was introduced to Gibraltar several hundred years ago and is the only nonhuman primate living "wild" in Europe. Today, the Barbary macaque is cared for and maintained by the British army stationed on Gibraltar.

BABOONS, MANDRILLS, AND DRILLS   The baboons and mandrills are large monkeys displaying marked size differences between the sexes. They have long muzzles possessing extremely long canine teeth, particularly in the males (see fig. 5.8). It is this appearance that has entitled them to the name "dogheaded" baboons, an appellation that has been used since Aristotle. They are well adapted for ground living, although they retreat to the trees or, in the case of the gelada and hamadryas ba-

2.28 *Papio hamadryas*, the hamadryas baboon of Ethiopa. These baboons were worshiped and mummified by the ancient Egyptians (see chap. 1). (Photo courtesy of Robert W. Sussman)

boons (fig. 2.28), to rocky cliffs at night. Most baboons have a catholic diet consisting of fruits, flowers, buds, roots, tubers, and a wide range of invertebrates. In addition, they will kill and eat hares, birds, young antelopes, and baby vervet monkeys. The savannah baboon (*Papio cynocephalus*) also practices cooperative hunting of small game by chasing prey toward another baboon in hiding, who, at the appropriate time, grabs the unsuspecting animal. The creature is quickly dispatched, after which the meat is shared with the other baboons (Kavanagh 1984).

The mandrills and drills (*Mandrillus sphinx* and *M. leucophaeus*) live in the forests of western Africa. Mandrills are very large, with the males getting to be from 45 to 50 kilograms (about 100 lb). Females are much smaller, as is generally true in all baboons. Drills are quite a bit smaller than mandrills. Male mandrills are noted for the prominent swellings along their long muzzles, which are colored with bright red and blue strips and assumed to be a sexual signal.

The gelada baboon (*Theropithecus gelada*, see fig. 4.8) from the plateaus of northern Ethiopia has been the sole subject of a book edited by

Jablonski (1993). The gelada baboon is the only survivor of a once widely distributed group of baboons comprising three species, with a fossil record going back at least 4 million years. Today the gelada represents a mixture of features that are probably related to its adaptation to the cold, dry environment of the Ethiopian high plateaus. They have an excellent opposable thumb, which they use quite efficiently in plucking up the blades of grass, complete with roots, which form the bulk of their diet. Their high-crowned molars are also well adapted for eating grasses.

MANGABEYS   The mangabeys (*Cercocebus*) live in the rain forests of central and western Africa. They are medium size, weighing from 8 to 10 kilograms, with the males somewhat larger than the females. They are arboreal monkeys occupying the high levels of the forests. The four species are mainly frugivorous (fruit-eating), although flowers, bark, and invertebrates are included in their diets.

GUENONS   Guenons make up a large group of small monkeys living in the forests of much of sub-Saharan Africa, where they have adapted to a variety of ecological niches. They are essentially arboreal monkeys, but some species, particularly *Cercopithecus aethiops*, the savannah monkey, have become quite terrestrial. Their locomotion is quadrupedal on the ground or in the trees, and they use their long tails as balancing organs when they leap through the trees. They are the most common of African monkeys and belong to the genus *Cercopithecus*, which refers to their long tails (*cerco* means tail in Greek and is the prefix for the family name, Cercopithecidae). Guenons range in size from small to medium, with moderate sexual dimorphism. They are essentially omnivorous, eating a variety of foods, which include both vegetable and animal products.

TALAPOINS   The smallest Old World monkey is the talapoin (*Miopithecus*), which weighs only about 1.3 to 1.8 kilograms (about 2–3 lb) when adult, with females somewhat smaller than males. They inhabit portions of the western African country of Gabon, preferring streams and swamp forests. They are excellent swimmers and use this ability to escape predators. In ecological adaptation and appearance they resemble the squirrel monkeys of South America.

PATAS MONKEYS   The quadrupedal patas monkey (*Erythrocebus*) is sleek of body and has long legs. It is the prime speedster among primates, capable of attaining speeds of 35 miles per hour (Kingdon 1971). Their distribution is in the open-wooded steppes and savannahs south of the Sahara in Africa. Their genus name refers to their predominant color of reddish hair, which is highlighted by a white moustache.

Apes

The apes are divided into two groups: the gibbons, or lesser apes (family Hylobatidae) and the great apes (chimpanzees, gorillas, and orangutans, family Pongidae). Of the three genera of great apes, two live in Africa (the chimpanzee, *Pan*, and the gorilla, *Gorilla*) and one in Southeast Asia (the orangutan, *Pongo*). In contrast to most prosimians and all but two species of monkeys, apes lack tails.

GIBBONS   According to most primatologists there is one genus of gibbons, *Hylobates*, with several species living in Southeast Asia. Some authorities, however, believe the larger gibbons, the siamangs of Sumatra and the Malay peninsula, belong in their own genus, *Symphalangus*. In the classification presented here, both gibbons and siamangs are treated as one genus, *Hylobates*. Gibbons are the acrobats of the primates. They are accomplished brachiators (arm-swingers) who prefer the closed canopies of the rain forests (fig. 2.29). Males and females are about the same size, roughly between 5 and 7 kilograms (11–16 lb). Siamangs are somewhat larger, 10 to 12 kilograms (22–27 lb). Both groups live in monogamous families (a male, his female mate, and two to four offspring). As the offspring become sexually active they leave the family unit to form their own families.

ORANGUTANS   The orangutans (fig. 2.30) are found only on the islands of Borneo and Sumatra today, but were, until rather recently, spread throughout southern Asia including much of China. Most primatologists classify them as one species living on both islands, although others believe that there are two subspecies, *P. pygmaeus pygmaeus* on Borneo and *P. pygmaeus abelli* on Sumatra. They are creatures of the rain forest, spending much of their time in the trees suspended by their long arms as they slowly progress through the dense forests. They do not brachiate. Orangutans are arboreal, rarely coming to the ground, but when they do, they walk quadrupedally using their fists for support. As adults, the males weight between 45 and 100 kilograms (100–220 lb),

2.29 The white-handed gibbon from Southeast Asia, *Hylobates lar*. Gibbons are the most accomplished of the brachiators. (Redrawn from Few 1991)

while the females weigh 35 to 50 kilograms (75−110 lb). In addition, adult males develop large fatty pads along the sides of their faces as well as large throat pouches housing laryngeal air-sacs, which may extend into their armpits (Winkler 1986). Orangutans are primarily fruit eaters but also consume ants, termites, and bees. Apparently, they rarely eat meat.

GORILLAS   Gorillas live in the forests of western and central Africa and are separated into the western and eastern lowland gorillas and the eastern mountain gorilla inhabiting the Virunga range of extinct volcanic mountains of Uganda and Rwanda. Gorillas are generally considered to constitute a single species (*Gorilla gorilla*) with three subspecies.

2.30 The shaggy red orangutan, *Pongo pygmaeus*, is found today only on Borneo and Sumatra. Notice the close-set eyes characteristic of the orang. The large layer of fat and skin surrounding the sides of the face and neck indicates that this is a male. (Redrawn from Few 1991)

The gorilla is the largest of the great apes, with wild males weighing between 155 and 165 kilograms (340–370 lb). Females are generally about half the size of males. In captivity, males may even get heavier: "Ngagi," a male eastern lowland gorilla in the San Diego Zoo, weighed 636 pounds (Napier and Napier 1967).

The mountain gorilla (*Gorilla gorilla beringei*) normally lives between about 2,800 and 3,400 meters (9,000–11,000 ft) above sea level and is the most endangered of the three subspecies of gorilla (fig. 2.31 a and b; see chap. 11). Gorillas are vegetarians, eating a wide variety of

2.31 a) Mountain gorilla, *Gorilla gorilla beringei*, mother holding her infant. b) Two mountain gorillas resting between meals. (Photos courtesy of Linda E. Duchin)

2.32 *Pan paniscus*, the pygmy chimpanzee, is also known as bonobo. A mother with an infant riding on her back. Also, note the knuckle-walking position of the mother's right hand. (Photo courtesy of Ellen Ingmanson)

leaves, stems, bark, roots, vines, and wild celery. The mountain gorilla consumes large quantities of bamboo, while the lowland gorillas eat large amounts of fruit. There is little evidence that gorillas eat meat.

CHIMPANZEES   The remaining great apes, *Pan troglodytes*, the "common" chimpanzee, and *Pan paniscus*, bonobo, inhabit the forest regions of western Africa. Bonobo (fig. 2.32) is regarded by some primatologists as a subspecies of *P. troglodytes* but is here treated as a separate species. Its distribution is restricted to central and western Zaire between the Congo River on the west and the Lualaba River on the east.

P. *troglodytes*, on the other hand, has three subspecies that inhabit western and central Africa from southeastern Senegal to western Tanzania and Uganda (fig. 2.33). Chimpanzees are arboreal but spend much time on the ground. When on the ground they walk quadrupedally and support their weight on their knuckles like gorillas (see chap. 7). Males weigh about 40 kilograms (88 lb), while females weigh 30 kilograms (66 lb). It is interesting to note that the pygmy chimpanzee is not much smaller than the common chimpanzee. Chimpanzees eat large amounts of fruit in addition to leaves, seeds, and many invertebrates. They also

2.33 The common chimpanzee, *Pan troglodytes*, is the most common of the two species. As with the bonobo and gorilla, it is a knuckle-walker when on the ground. (Redrawn from Few 1991)

eat meat in the form of several species of antelope, bush pigs, birds, and even other primates, such as young colobus monkeys and baboons. Knowledge of chimpanzee behavior has been greatly increased over the past three decades by the pioneering work of Jane Goodall (Goodall 1986 and 1990, and see chap. 9).

3

Blood Groups,

Chromosomes,

and DNA

BLOOD GROUPS

Landsteiner's discovery in 1901 of blood groups in humans initiated what turned out to be one of the most active and important research areas in clinical medicine—human genetics—and, today, organ transplants. As early as 1925 it was known that the red blood cells of non-human primates had antigens (substances that stimulate the production of antibodies) similar to those of humans. Then, in 1940, Landsteiner and Weiner (1940) discovered the Rh factor in rhesus monkeys, one of the most important discoveries in the history of blood group studies in primates (see below). This discovery of a "monkey-type" antigen in human blood was a great impetus to study the blood of other nonhuman primates using other human antigens. Later, antisera were made from the red cells of nonhuman primates themselves and then used to ascertain the specificities of the blood of other nonhuman primates. The former are known as human-type blood groups, while the latter are simian-type blood groups. Today we know that there is no clear-cut separation between these two categories, since simian-type blood groups are found in humans (Socha and Moor-Jankowski 1979). Information regarding the blood groups of nonhuman primates is much more limited than it is for humans. At present there are several separate blood group systems under study in nonhuman primates, such as the well-known ABO, M-N, Rh, Lewis, and Duffy (Socha et al. 1995). There are blood group data for all of the anthropoid apes, some species of Old World monkeys, very few species of New World monkeys, and only a few prosimians. The ABO blood group system is well documented in nonhuman primates, and our discussion will pertain mainly to this system, with a few comments regarding the Rh system (for a recent review of nonhuman primate blood group systems, see Socha et al. 1995).

The ABO Blood Group System

The ABO group was the first blood group system discovered in humans by Landsteiner in 1901. Some years later it was identified in *Pan*, and today it has been found in all primates that have been examined. Primates have four blood groups: A, B, O, and AB (table 3.1). There are two agglutinable antigens on the red cells, A and B, and two anti-A and anti-B substances present in the sera which determine an animal's blood type. Thus, red cells of a group A animal contain A but not B agglutinogen (antigens) while the serum has anti-B but not anti-A agglutinins. If a transfusion is necessary it should be from an individual whose blood cells do not become agglutinated when mixed with the blood of the recipient. In humans, persons with group O are considered universal donors while persons with the AB group are universal recipients (table 3.2). These substances are present not only in primate blood but in their saliva as well, since the ability to secrete the A, B, and O antigens in the saliva is inherited as a dominant character. Also, since A and B substances are not normally detectable on the red cells of primates, other than in hominoids, saliva is used for identifying the ABO groups in other primates (Socha 1986 and Socha et al. 1995).

The ABO blood group frequencies for the anthropoid apes are presented in table 3.3. The two species of gorillas and the siamang have only B, while the pygmy chimpanzee has only the A group. These three taxa are monomorphic (one form) with respect to the ABO system, while the other apes are polymorphic (many forms). It is interesting

Table 3.1   The Four ABO Blood Groups
of Human and Nonhuman Primates

Blood Group	Red Cell Antigens	Serum
O	None	Anti-A and Anti-B
A	A	Anti-B
B	B	Anti-A
AB	A and B	None

Table 3.2   Reactions in the A-B-O Blood Groups

Serum	Red Cells			
	O	A	B	AB
O	–	+	+	+
A	–	–	+	+
B	–	+	–	+
AB	–	–	–	–

to note that the common chimpanzee is the only great ape having the O blood group.

The distribution of ABO blood groups among the Old World monkeys is presented in table 3.4. Of the twenty-one species appearing in table 3.4, only six are monomorphic; all others are polymorphic. A few New World monkeys and prosimians have been tested for the ABO blood groups as shown in table 3.5. Two species of New World monkeys are monomorphic, while both of the prosimians are monomorphic.

Besides the obvious importance of blood groups in clinical situations (blood transfusions and cases of materno-fetal incompatibility) they have also been of value in estimating times of evolutionary divergence, as well as in taxonomic classifications and as genetic markers. For example, the fact that the ABO antigens are present in all primates studied to date would suggest that ABO polymorphism was present in the

Table 3.3   The ABO Blood Groups of the Lesser and Great Apes

Animal	Blood Group (%)				Sample
	O	A	B	AB	
Pan troglodytes	10.1	89.9	0.0	0.0	972
Pan paniscus	0.0	100.0	0.0	0.0	14
Gorilla g. g.	0.0	0.0	100.0	0.0	50
Gorilla g b.	0.0	0.0	100.0	0.0	4
Pongo p.	0.0	56.1	16.9	27.0	91
Hylobates	0.0	18.9	41.9	39.2	143
Symphalangus	0.0	0.0	100.0	0.0	2

Adapted from Socha et al. 1995.

primate lineage prior to the separation of the Old and New World monkeys (Socha et al. 1995). Another important contribution has been in the study of hybridization of monkey populations living in overlapping geographical areas (Brett et al. 1976 and Socha et al. 1977).

The Rh Blood Group System

The Rh system is of interest to primate biologists because of its importance in evolutionary genetics and its practical implications for understanding the Rh factor on human red cells. As mentioned previously, the Rh factor was originally found in the rhesus monkey (*Macaca mulatta*), thus the Rh symbol refers to the rhesus monkey.

In humans there is a condition known as hemolytic disease of the newborn (*erythroblastosis fetalis*) which causes the oxygen-carrying red blood cells of the fetus to be destroyed. The result is usually death of the fetus. This rather rare situation (about 85 percent of humans are Rh positive while 15 percent are Rh negative) occurs when the parents have incompatible Rh factors. For example, if the mother is Rh negative and the father is Rh positive, then the fetus is Rh positive. The hemolytic disease is due to the reaction of the antibodies stimulated within an Rh-negative mother by the Rh-positive cells of the fetus. In subsequent

Table 3.4   The ABO Blood Groups of Old World Monkeys

Animal	Blood Group (%)				Sample
	O	A	B	AB	
Papio anubis	0.0	14.5	47.5	38.0	750
P. hamadryus	0.0	8.7	62.2	29.1	172
anubis/hamadryus hybrids	0.0	3.9	62.8	33.3	129
P. cynocephalus	0.0	31.0	33.0	36.0	80
P. ursinus	0.0	22.0	43.5	34.5	168
P. papio	1.0	14.4	49.5	35.1	188
Papio (species unknown)	0.5	24.5	38.0	37.0	184
Theropithecus gelada	100.0	0.0	0.0	0.0	20
Macaca arctoides	0.0	0.0	100.0	0.0	41
M. fascicularis	1.4	26.7	37.9	34.0	985
M. fuscata	0.0	0.0	100.0	0.0	14
M. mulatta	1.0	1.0	97.0	1.0	215
M. nemestrina	74.2	15.7	7.9	2.2	89
M. radiata	0.0	41.5	36.4	22.0	77
M. sylvanus	0.0	100.0	0.0	0.0	32
Mandrillus	0.0	100.0	0.0	0.0	4
Cercopithecus (vervet monkeys)	0.5	70.1	9.4	20.0	562
Presbytes entellus	0.0	0.0	88.9	11.1	18
Cynopithecus	3.8	88.5	7.7	0.0	26
Erythrocebus	0.0	100.0	0.0	0.0	26

Adapted from Socha et al. 1995.

Table 3.5   The ABO Blood Groups of New World Monkeys and Prosimians

Animal	Blood Group (%)				Sample
	O	A	B	AB	
Ateles (various species)	6.0	67.0	27.0	0.0	15
Saimiri sciureus	18.0	55.0	0.0	27.0	11
Cebus (various species)	1.0	56.0	33.0	0.0	9
Alouatta	0.0	0.0	100.0	0.0	52
Callithrix (various species)	0.0	100.0	0.0	0.0	45
Galago crassicaudatus	0.0	100.0	0.0	0.0	17
Lemur catta	0.0	0.0	100.0	0.0	13

Adapted from Socha et al. 1995

pregnancies these anti-Rh-positive antibodies from the mother enter the fetal circulation and destroy the fetal red cells, resulting in a severe anemia.

It is interesting to note that, among the anthropoid apes, only the red cells of the orangutan fail to react (no homologues) with the human Rh factor (Socha 1986 and Socha et al. 1995).

CHROMOSOMES

Chromosomes (literally, colored bodies) are present in each cell of an organism and carry the genetic information, or DNA (see below), which is arranged within each chromosome as genes. The study of chromosomes, known as cytogenetics, has become an important branch of molecular biology. Chromosomes can be compared among the various primate species and their phylogeny and evolution investigated.

Chromosomes appear as paired structures consisting of two identical chromatids joined to each other at the primary constriction by the centromere and have different morphological characteristics that can be used in their identification (fig. 3.1). A chromosome is metameric if the centromere is near the middle of the chromatids (fig. 3.1, M). If the centromere is near the end of the chromatids it is acrocentric (fig. 3.1, A). If the centromere is at the end of the chromatids it is telocentric (fig. 3.1, T). When the centromere is between the middle and terminal positions it is subterminal (fig. 3.1, S). Chromosomes are counted and studied during cell division, especially during metaphase, for it is during this time that the chromosomes separate and become most prominent as they line up along the plane of cell division.

The number and type of chromosomes in an organism's cell constitute its karyotype. A specific karyotype is characteristic for a species— for example, humans have 46 chromosomes (known as the diploid or 2N number) consisting of 23 pairs of chromosomes, of which, 22 are autosomes (any chromosome except a sex chromosome) and 1 pair are sex cells (XX, female or XY, male). The number of chromosomes present in the human germ cell (egg or sperm) is one-half the diploid number, or 23, and is known as the haploid (N) number of chromosomes. Karyotypes vary a great deal among living primates (2N ranges from 20 to 80), which indicates much selection during the course of primate history.

As new techniques have been developed using more sensitive stains, fluorescent dyes, and various banding methods, there has been a marked increase in our knowledge of primate karyotypes during the last decade. For example, G-banding of chromosomes allows the identification of each chromosome and any structural rearrangements that have occurred, while C-banding stains small regions of certain chromosomes and may be variable within and between species.

Changes in the number or shape of a chromosome (chromosal mutations) result in the rearrangement of genetic material, which is often fatal or deleterious to the organism. Chromosomal rearrangements include loss of chromosome material (deletion), duplication of material, turning pieces of chromosomes upside down (inversion), and translocation (exchange of material between chromosomes). Since such changes are rare, and a given number and form of chromosomes remain stable for a particular species, there is probably strong selection pressure against such events. However, these events do occur in populations, and if they take place over many generations the result is chromosome evolution. In other words, these rearrangements constitute the raw materials for karyotype evolution which may result in different species.

The majority of primates have been karyotyped, and there are various publications with lists and tables of primate karyotypes (Chiarelli

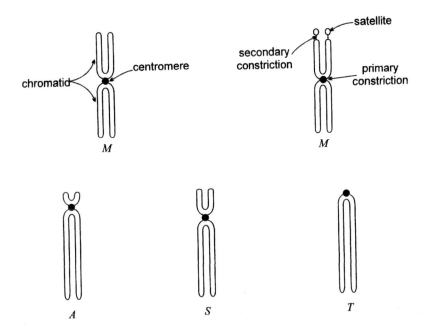

3.1 Morphological characteristics of chromosomes according to the position of the centromere (see text for discussion).

et al. 1979 and Ankel-Simons 1983). The number of chromosomes (2N) varies among primates from 20 to 80, which, incidentally, is also the range in prosimians. For example, some species of *Lepilemur* have the lowest number of chromosomes recorded among primates (2N=20), while the specialized little tarsiers have the most (2N=80). A recent cytogenetic study of the systematics of the Cheirogaleidae has suggested that all Malagasy primates are members of a monophyletic lineage that excludes the Afro-Asian lorisiforms (Yoder 1994). There are five genera in the Cheirogaleidae (*Cheirogaleus*, *Microcebus*, *Mirza*, *Allocebus*, and *Phaner*), and all have 2N=66 chromosomes except *Phaner*, which possesses a diploid number of 46. Because of this different karyotype, it has been suggested that the Cheirogaleidae be divided into two subfamilies: Cheirogaleinae and Phanerinae (Rumpler et al. 1995).

Among New World monkeys chromosome numbers range from 20 to 62. The lowest number is found in the titi monkey, *Callicebus torquatus*, while the highest number is in the woolly monkey, *Lagothrix*. Another species of titi, *C. moloch*, however, has 50 chromosomes (Bernischke and Bogart, 1976). This dissimilar chromosome number suggests that there has been a great deal of chromosomal evolution during the evolution of the Callicebinae. On the other hand, members of the family Callitrichidae are rather consistent in having 2N=44, 46, and (if Callimico is included as most authorities do today) 48 (but see Hershkovitz 1977). The Cebidae are much more variable, ranging from the lowest number 20 to the highest 62. The genus Cebus is rather consistent with 2N of 52 and 54, while Ateles has 34, Alouatta 52, and Saimiri 44 (Chiarelli, et al. 1979).

The 2N chromosome number of Old World monkeys varies from 42 to 72. The three types of chromosomes (metacentric, acrocentric, and submetacentric) are present in all species of Cercopithecus. The two subfamilies of Old World monkeys have quite different diploid ranges: the Cercopithecinae vary from 42 to 72, while the Colobinae possess a more restricted range from 44 to 48 (Marks 1994). Within the Cercopithecinae the vervet and guenon monkeys (Cercopithecini) 2N numbers vary from 48 to 72, while the macaques and baboons (Papionini) all have a 2N of 42 (Dutrillaux et al. 1986 and Marks 1994). It has been suggested that the evolution of chromosomes in the Cercopithecinae has followed a course of decrease in size and an increase in number as a consequence of recurrent fissions of the largest chromosomes (Dutrillaux et al. 1986). Primate species having large diploid numbers, as in Cercopithecus mitis (2N=72), will also have an increase in recombinations by crossovers during meiosis (reduction of diploid number to haploid number) compared to animals with lower diploid numbers, for example, Papionini (2N=42). This means that linkage groups are "much more frequently disrupted" in animals with higher diploid numbers, which can have important evolutionary consequences (Dutrillaux et al. 1986).

The chromosome number ranges from 44 to 52 in hominoids. According to Marks (1994) the highly derived chromosomes of gibbons (2N=38 to 52) have undergone rapid and extensive changes since separating from the other hominoids. Gibbons have mostly metacentric and submetacentric chromosomes, although Symphalangus has a single pair of acrocentric chromosomes (Ankel-Simons 1983). The great apes have 2N=48 chromosomes, and all have the three types of chromosomes. The human 2N=46 has resulted from the fusion of two acrocentric chromosomes to form the number 2 chromosome of humans. Humans also have three types of chromosomes. The sex chromosomes (X and Y) are similar in the great apes and humans—that is, the X chromosome is submetacentric and the Y chromosome is acrocentric.

The 2N range of primate chromosome numbers is represented by the prosimians (20 to 80), while the great apes and humans have the smallest range of chromosomes (46–48). Different species may have the same number of chromosomes, but their form may be quite different, so the characterization of a species's karyotype must consider both the number and form of the chromosomes. Also, chromosomal polymorphisms (alternative forms) are rather rare among primates. For example, some primate species are almost monomorphic (*Papio cynocephalus*) while different populations of the Colombian owl monkey, *Aotus trivirgatus*, may have several different karyotypes, diploid numbers of 52, 53, and 54 (Marks 1994 and Martin 1990). Chromosomal polymorphisms obviously occur in primate species and provide an important source of "raw material for evolutionary change in the karyotype" (Martin 1990, p. 113).

A symposium titled "Molecules Versus Morphology in Primate Systematics" (Cartmill and Yoder 1994) was concerned with issues, problems, and questions regarding the relative merits of molecules versus structural morphology for phylogenetic reconstruction. The debate has been going on for years, and, although there is no clear consensus, it is generally felt that all relevant information is important and should be used when looking for phylogenetic history. Perhaps Disotell (1994, p. 53) stated it as clearly as any in his analysis of papionin phylogeny when he wrote, "Questions about the inherent superiority of molecular or morphological data are based upon a false dichotomy. Rather, both types of data should be used in an attempt to corroborate the different phylogenetic hypotheses generated by their respective analyses." Such scientific gatherings are necessary from time to time in order to introduce fresh ideas, present different view points, and discuss diverse methods and analyses regarding issues of mutual concern.

DNA

One of the most important biological discoveries of the twentieth century was the identification of the structure and function of deoxyribonucleic acid, DNA (Watson and Crick 1953). This material, which with protein makes up the majority of the cell nucleus, contains the genetic message that enables the cell to replicate itself, as well as providing information necessary for the cell's many other functions. Today we know that DNA is the basic copying material for essentially all life on Earth, from fungus to humans. As Gribbin pointed out (1993, p. 80), "nothing *wants* to evolve." The essential process of life, therefore, is replication, and it is important that the copying process proceeds as accurately as possible. From time to time, however, mistakes (mutations) do occur in the copying process, but these are usually harmful to the organism. Occasionally a mutation happens that is beneficial to the organism, selection occurs, and the mutation survives and spreads through the population. The result is evolution by natural selection which through millions of years has produced the diversity of life present today on Earth.

One consequence of DNA studies has been the development of DNA systematics. Comparative molecular biologists, who include primatologists and anthropologists, have built up over the past several decades a large number of DNA sequences for many species of primates. These studies have made the primates one of the best-known groups of higher organisms (Miyamoto and Goodman 1990). Indeed, analysis of this excellent data base has not only provided much important information regarding the evolution of DNA sequences, but has also supplied valuable data concerning the taxonomy of living primates, as well as their phylogenetic history.

Advances in DNA technology have made it possible to sequence genes and compare them with genes from other species. Also, fragments of DNA from two different species can be combined into a hybrid segment, a process called DNA hybridization. There are four types of bases in DNA; adenine, guanine, cytosine, and thymine. The number of matches and mismatches between the bases on the hybrid strands of each DNA fragment allows the determination of the degree of similarity between the two species at the DNA level.

Such studies have provided information on the degree of genetic relationships among primates and have demonstrated that the DNA of chimpanzees, gorillas, and humans differs by only about 2 percent. This means that when strands of DNA from any two of these animals are combined about 98 percent of the bases match. Humans differ from orangutans by about 4 percent and from baboons by about 8 percent. In fact, Dr. Morris Goodman, a doyen of primate molecular biology, has suggested that chimpanzees and humans be placed in the same subtribe, Hominina, while the gorillas have their own subtribe, Gorillina, and that both subtribes be placed in a single tribe, Hominini (Goodman et al. 1989). The orangutans and gibbons have their own subfamilies, and all of these taxa are placed in the family Hominidae. A molecular classification of primates is presented in table 3.6. This classification departs somewhat from the morphological one presented in chapter two since opinions differ regarding relationships among some of the taxa.

Table 3.6   Molecular Classification of Living Primates

Order Primates
  Suborder Strepsirhini
    Superfamily Lemuroidea
      [lemurs and dwarf lemurs]
    Superfamily Lorisoidea
      [galago and lorises]
  Suborder Haplorhini
    Semisuborder Tarsiiformes
      [tarsiers]
    Semisuborder Anthropoidea
      Infraorder Platyrrhini
        [New World monkeys]
      Infraorder Catarrhini
      Superfamily Cercopithecoidea
        [Old World monkeys]
      Superfamily Hominoidea
        Family Hominidae
        Subfamily Hylobatinae
        [gibbons]
        Subfamily Homininae
          Tribe Pongini [orangutans]
        Tribe Hominini
          Subtribe Gorillina
            [gorillas]
          Subtribe Hominina
            [humans and chimpanzees]

Modified from Goodman et al. 1990.

For example, tarsiers have been traditionally placed with the suborder Prosimii as an infraorder, Tarsiiformes, even though they possess many characteristics in common with members of the suborder Anthropoidea (chap. 2). Many authorities, however, believe that living primates are better classified as two suborders, Strepsirhini and Haplorhini (table 3.6). In this system tarsiers are considered members of the Haplorhini, and lemurs and lorises are the only members of the Strepsirhini.

Perhaps a more contentious issue between the molecular and morphological approaches to primate taxonomy is the relationships among the great apes and humans. As just mentioned, molecular biologists tend to subdivide the hominoids into more subdivisions within the superfamily Hominoidea (table 3.6) than do morphologists (chap. 2). A recent paper (Goodman et al. 1994, p. 3), using sequence data from mitochondrial (a cellular organelle involved in energy metabolism) and nuclear genomes, stated that "humans and chimpanzees share the longest common ancestry." There is little doubt today that the chimpanzee is the human being's closest relative; the only question appears to be the closeness of the relationship.

Molecular clocks have also been established using molecular data. Two types of clock have been recognized: global and local. Both are based on the assumption of constant rates of change at the nucleotide level. In other words, if the differences between two taxa are the result of mutations that have occurred at the same rate, it should be possible to calculate the amount of time it took to accumulate the present level of mutations. We know, however, that rates have changed through time among various primate taxa and that there has been an appreciable slowdown from lower primates to anthropoids, from anthropoids to hominoids, and again among hominoids. Indeed, "the rate of nucleotide evolution in humans appears to be slower than that for other higher primates" (Miyamoto and Goodman 1990, p. 208). This overall rate decrease in higher primates which culminates in humans has prompted Goodman (1985, p. 11) to refer to it as "the hominoid slowdown." There have been several suggestions to explain the different rates of molecular evolution among primate lineages. For example, decreases in evolutionary rates may be the result of "a general increase in generation time in hominoid species," or they may be due to "an early accumulation of selectively neutral or advantageous mutations, while most later mutations were harmful" (Porter et al. 1995, p. 54). The jury is still out on definitive answers to these observations, and the solution may turn out to be a combination of these theories; certainly, differential selective forces have operated throughout primate phylogeny.

Molecular clocks have their critics, and there has been much heated debate over the past several years regarding their accuracy. If average mutation rates are variable through time, that is, higher or lower than presumed, the clock may be inaccurate. There is no way of ever knowing what the average primate DNA mutation rate was 30 or 40 million years ago (MYA). Certainly, molecular clocks should be used with caution, and special clock calculations have been used to compensate for unusual rates (Sarich and Cronin 1976). There is agreement, however, that the similarity of DNA sequences among primate taxa indicates their degree of biological relationship; it is the timing or the divergence dates of these events that cause the problems.

4

The Skull

The study of bones is called osteology. Bones are the hard, dry, often brittle structures we see displayed in museums as skeletons representing the last remains of some recent or fossil animal. Osteology is more than just putting skeletons back together, although this is important. Bones have many functions as part of an animal's skeleton: they are responsible for protecting soft tissues (the brain, for example); for providing the rigidity and framework for a series of mechanical levers to which muscles and ligaments attach for support and locomotion (see chap. 7); for serving as a reservoir for minerals (most of an animal's total body calcium is stored in bones); and for the production of blood cells. Living bone is a vital, dynamic tissue that is malleable. Therefore its structure can be changed, bent or straightened, for example, by outside forces during the life of an animal. Bone is also continually undergoing internal reconstruction or remodeling. Remodeling is the process by which mature bone is destroyed and gradually replaced by new bone. This pro-

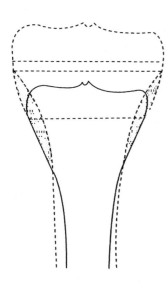

4.1 Diagram of the growing end of a long bone (tibia) showing the process of modeling resorption (stippled area) required to maintain the normal shape of the growing shaft. (Redrawn from Weinmann and Sicher 1947)

cess occurs throughout life and in all bones of the body. Also, modeling resorption is necessary to maintain the normal shape of a growing bone as exemplified by one of the bones of the leg (the tibia) (fig. 4.1). In this case, the amount of modeling resorption necessary to maintain the normal shape of the growing bone is shown in the stippled areas. Because of this remodeling process, bone has the ability to repair itself after a fracture or respond to the movement of teeth through the jaw. Without bone remodeling there could be no orthodontics. Bone even responds to the presence of soft tissues such as the brain and blood vessels. For example, the exact locations of the large arteries and veins on the surface of the brain (the middle meningeal vessels) are indicated as grooves on the inner surface of the cranial vault. Finally, next to the enamel of teeth, bone is the hardest and most compact tissue in the mammalian body. Consequently, bones and teeth are the structures most often preserved as fossils.

There are two types of bone development or ossification: intramembranous ossification and endochondral ossification. Intramembranous ossification occurs when bone develops within a connective tissue membrane and does not involve the removal of cartilage. Bones formed in this manner are mainly bones of the skull: frontal and parietal bones, bones of the upper face, parts of the sphenoid, occipital, and temporals, as well as most of the mandible.

Endochondral ossification occurs when ossifying tissue replaces a former cartilaginous model. This is the most common type of ossification and includes the bones of the axial and appendicular skeletons (with the exception of the clavicle; see chap. 7) and, in the skull, several bones of the nasal cavity as well as bones of the cranial base.

Figures 4.2 through 4.7 present a series of primate skulls in two views, front and side, for reference during the following anatomical discussion of primate skulls.

THE SKULL

The primate skull is a three-dimensional, rather spheroid structure and is the most complex organization of bones in the skeleton. This is due primarily to its adaptation to the brain and the organs of special sense. The skull, therefore, has several important functions: it encloses and protects the brain, it accommodates the special sense organs (vision, hearing, smell, and taste), it supports the dentition for obtaining and chewing food, and it is where the muscles of mastication (chewing) and facial expression (e.g., smiling or pouting) attach.

The size and proportions distinguishing the primate skull from other mammalian skulls illustrate a series of morphological modifications associated, at least initially, with an arboreal mode of life. These include an increase in visual acuity with larger and more frontally oriented orbits which are surrounded by a bony ring; a reduction in smell usually accompanied by reduction of the nasal region; a gradual enlargement of the brain with commensurate increase in the neurocranium (braincase); a development of various degrees of upright trunk positions, which changes the position of the foramen magnum (the hole at the base of the skull through which the spinal cord passes) from facing backwards (as in more pronograde or four-legged mammals) to a more forward position pointing downwards; and increased use of the forelimbs for prehension and procuring food, resulting in a general reduction of jaws and teeth.

It must be remembered that these anatomical features represent evolutionary adaptations which have undergone many changes during the evolution of primates and consequently are quite variable in living forms. Indeed, the structural, ecological, and behavioral diversity of living primates is one of their major characteristics. For example, a nocturnal primate such as *Tarsius* has extremely large bony orbits which house its large eyes for better night vision, while the gelada baboon (*Theropithecus*) has a prognathic muzzle due to the heavy jaws and large teeth associated with its specialized diet of grasses and rhizomes (fig. 4.8). Because of this sort of diversity, primates are more often characterized by their "adaptability" than by their "adaptations."

The skull is divided into two separate parts: the cranium and the mandible (lower jaw). The cranium can be further divided into the neurocranium, which houses and protects the brain, and the splanchnocranium, which forms the facial skeleton. These regions are somewhat arbitrary since they tend to overlap with each other, and it is difficult to draw absolute boundaries between them. A few of the bones forming the different parts of the skull appear singly; most, however, are in pairs. The mandible is separate and joins the cranium in the mandibular fossae (depression) of the temporal bones.

Bones of the Face

The facial region consists of those bones surrounding the mouth and nose and several bones of the orbit surrounding the eye (fig. 4.9). These are the paired nasal, maxilla, zygomatic, lacrimal, palatine, inferior nasal concha, and vomer, the unpaired mandible, and the horseshoe-shaped hyoid bone, which is the bony support of the tongue.

An important feature of the primate face is its angle with the basicranial axis of the skull (fig. 4.10). Typically, the mammalian face lies in front of the cranium, forming only a slight angle with the cranial base. The angle becomes more acute in primates, especially in the anthropoids, where the face is positioned beneath the cranium rather than in front of it. In figure 4.10 note the shift in the position of the foramen magnum from pointing backwards as it does in dogs to pointing more downwards and under the base of the skull in primates. The changes in the craniofacial relations are correlated with the greater cerebral development of primates, and the alteration in the plane of the foramen magnum is associated with the reduction of the face as well as with the orthograde or more erect posture of the trunk. The result has been that the head is balanced on the first cervical vertebra forming the atlanto-occipital joint. Along with these changes and a commensurate reduction of the olfactory areas, there has been a recession of the snout, although there is considerable variation in muzzle development among living primates and for various reasons (see figs. 4.2 to 4.7). In some species— for example, the more terrestrial cercopithecids, such as baboons and mandrills—long muzzles are associated with their large teeth and dietary habits. Napier and Napier (1967) have distinguished such "dental"

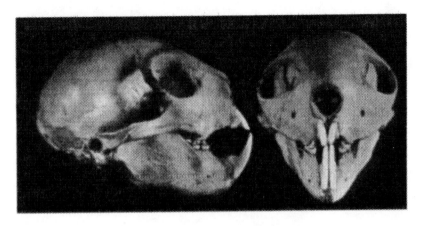

4.2 Side and front views of an aye-aye skull (*Daubentonia madagascariensis*).
(Elliot 1912)

muzzles found among certain ground-living groups (fig. 4.8, T. *gelada*) from other "olfactory" muzzles present in some Lemuridae and Lorisidae (figs. 4.3 and 4.4).

The frontal bone is low and sloping in many primates, but has become more vertical, conforming to the large frontal lobes of the brain during the later stages of primate evolution. The frontal bone develops as paired bones in all primates and remains separated by a midline (metopic) suture in many prosimians (fig. 4.4). In anthropoids the suture usually disappears before adulthood, but if it remains it is also called a metopic suture. The supraorbital ridges, or tori, are present to some degree in most primates and are particularly well developed in certain catarrhines. These ridges are usually better developed in the males of most species. Consequently they are one of several structures used to identify the sex of skulls.

The maxilla, or upper jaw, containing the upper teeth, is composed of two paired bones in nonhuman primates, the premaxilla and maxilla (fig. 4.9). Only the upper incisor teeth develop and occupy the premaxilla. The premaxilla is quite variable in size, extending upward for various distances between the nasal and maxillary bones. It is reduced in modern lemurs, and apparently the small size of the premaxilla is correlated with their diminutive incisors, which, in *Lepilemur*, are completely absent (Swindler 1976). The premaxilla is observable well into adult life in most primates, and only rarely (in 2 of 610 skulls) is there a premaxillary-frontal contact in modern primates (Swindler and Tarrant 1973). The premaxilla is present in humans but disappears early in

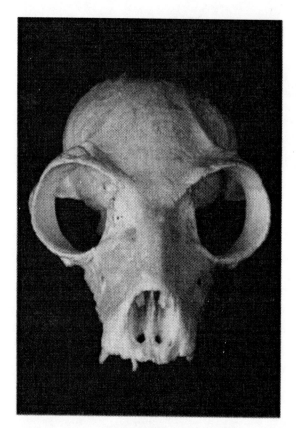

4.3 Skull of a ring-tailed lemur (*Lemur catta*); female. Frontal view × 1.2 and side view × 1.1. (Figs. 4.3 to 4.7 were photographed by Joseph R. Siebert, Department of Laboratories, Children's Hospital and Medical Center, Seattle)

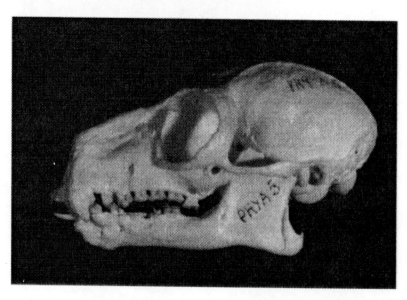

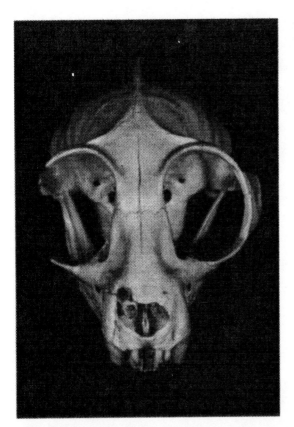

4.4 Skull of a large-eared greater bushbaby (*Otolemur crassicaudatus*); male. Frontal view × 1.5 and side view × 1.4.

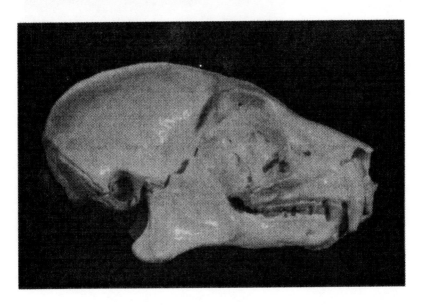

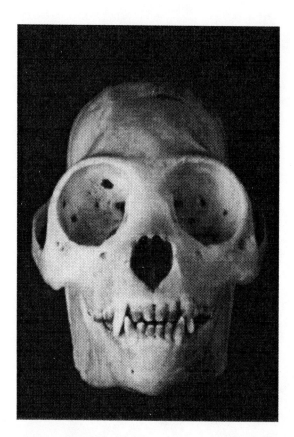

4.5 Skull of a Guatemalan howler monkey (*Alouatta villosa*); female. Frontal view × 0.9 and side view × 0.6.

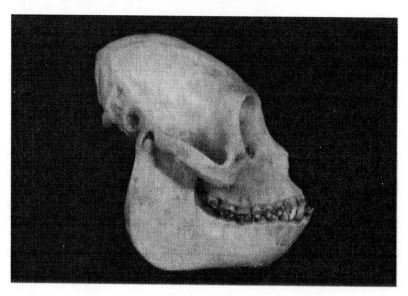

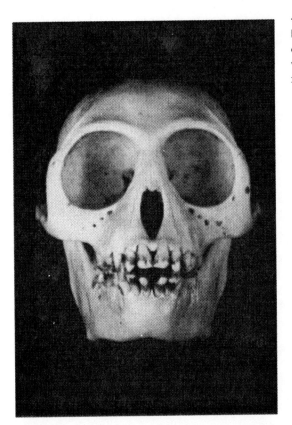

4.6 Skull of a silvered leaf monkey (*Presbytis cristata*); female. Frontal view × 1.0 and side view × 0.9.

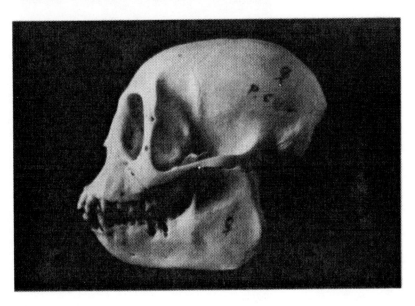

4.7 Skull of a common chimpanzee (*Pan troglodytes*); male. Frontal view × 0.6 and side view × 0.4.

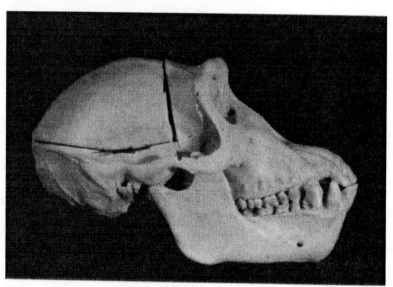

embryological development. A diastema (a space between two adjacent teeth) is frequently present between the upper lateral incisor (premaxilla) and canine (maxilla) (see fig. 4.14). Also, there may be a diastema between the lower lateral incisor and canine. Both diastemata are highly variable among primates, and Schultz (1963) showed that the maxillary diastemata were primarily the result of differential growth of the premaxillary bone and the age at which the premaxillary sutures close rather than being associated with the size of the canines.

The premaxilla has had an interesting history in comparative primate osteology since its discovery by the famous Greek physician Galen in the second century A.D. From that time until the appearance of Vesalius's *De Humani Corporis Fabrica* in 1543, it was assumed that humans possessed a separate premaxillary bone. Conducting his own dissections, Vesalius was able to demonstrate that Galen had indeed studied only monkeys after all, and had not demonstrated the presence of the premaxillary bone in humans.

The debate over the presence or absence of the human premaxillary bone in humans continued into the present century (Ashley-Montagu 1935). The reason for this long debate is significant. If humans lacked the premaxillary element, it meant that they differed radically from other primates and this feature could be used to separate *Homo sapiens* from all other primates.

Today we know that the premaxilla does exist initially in the human embryo, but that it disappears early in embryological development by fusing with the maxilla on the anterior surface of the facial skeleton. This debate, however, illustrates how a rather detailed bit of comparative osteological information can be used to point out the significance of ontogenetic (growth) information in the study of primate evolution (see chap. 8).

Included in the category of bones of the face are three small bones referred to as ear ossicles, located inside the middle ear cavity of the temporal bone of mammals. These conduct sound waves from the tym-. panic membrane (the ear drum) to the deepest recesses of the inner ear, where they are picked up by the auditory nerve. The ear ossicles are the malleus, incus, and stapes, which have had an interesting phylogenetic history among vertebrate animals. The stapes is homologous to the columella (the single ear bone) of many vertebrates (amphibians and reptiles), while the malleus and incus are homologous to the articular and quadrate bones in the lower jaws of these animals. Figure 4.11 depicts this important chapter of mammalian evolution from bony fishes to

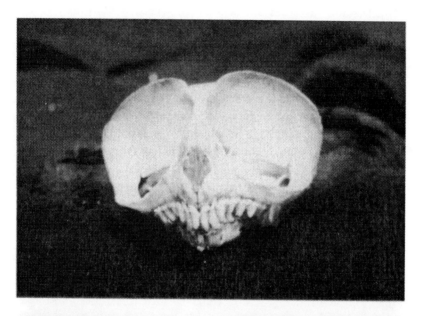

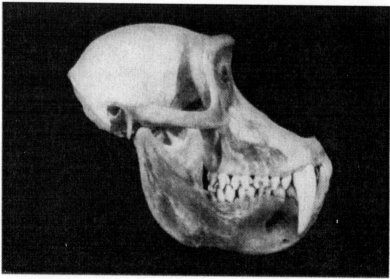

4.8 Top: Front view of a male tarsier skull (*Tarsius bancanus*). (Photo by author) Bottom: Side view of a male gelada skull (*Theropithecus gelada*). (Photo courtesy of Nina G. Jablonski)

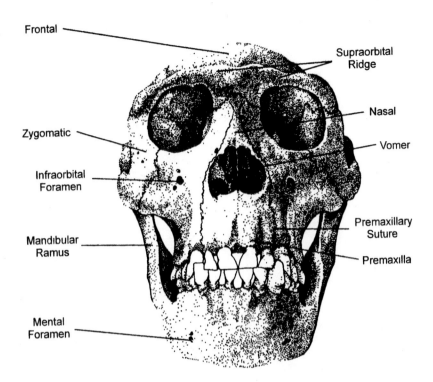

Frontal

Supraorbital
Ridge

Nasal

Zygomatic

Vomer

Infraorbital
Foramen

Mandibular
Ramus

Premaxillary
Suture

Premaxilla

Mental
Foramen

4.9 Front view of a male chimpanzee skull. (Redrawn from Swindler and Wood 1973)

mammals as it traces the changes of structure and function of the three
bones of the mammalian middle ear. It has been suggested by paleontologist Timothy Rowe that the malleus and incus lost their hinge function of attaching the lower jaw to the skull as the early mammalian
neocortex portion of the brain expanded (see chap. 6), which in turn
freed these bones, allowing their migration back along the skull (Fischman 1995). The result was the evolution of a delicate sound-amplify-
ing system known as the middle ear, which is found only in mammals.

Bones of the Orbit

The paired lacrimal bones and lacrimal canals are positioned outside the
orbits in lemuriforms and lorisiforms (figs. 4.3 and 4.4) but are located
within the orbits in the Anthropoidea, though occasionally the lacrimal
canal may be found on the orbital margin or slightly outside the orbit
in some anthropoid genera. The topographic relations of the bones of
the medial wall of the orbit are shown in figure 4.12. The typical mam-

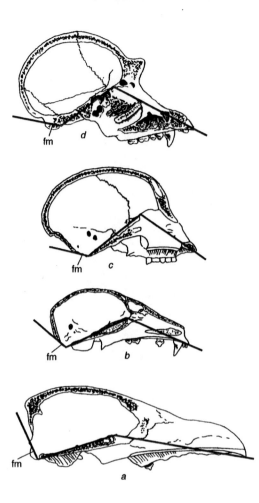

4.10 Sagittal view of skulls showing the relationship of the facial skeleton to the basicranial axis and the change in the position of the foramen magnum (fm): a) *Canis familiaris* (dog); b) *Lemur catta*; c) *Cebus*; d) *Pan troglodytes*. (a, b, and c adapted from Hill 1953; d, from author's collection)

malian condition of a wide contact between the frontal bone and the maxilla is represented in (a), while in (b) the palatine articulates with the lacrimal in modern lemuriforms, but the ethmoid contacts the lacrimal and also intervenes between the frontal bone and the maxilla in lorisiforms, platyrrhines, and catarrhines (c). Commensurate with the larger, more frontally oriented eyes, there has been a change in the organization of the bones on the lateral and posterior regions of the orbit. This may have been in response to forces directed to the orbital region during chewing, or as extra bony support for the eyeball as it became more frontally directed. Thus, a strut or ledge of bone has developed in primates to connect the frontal bone to the zygomatic bone. It is known as the postorbital bar (fig. 4.13). The postorbital bar is present in all living primates as well as in some other mammals, for ex-

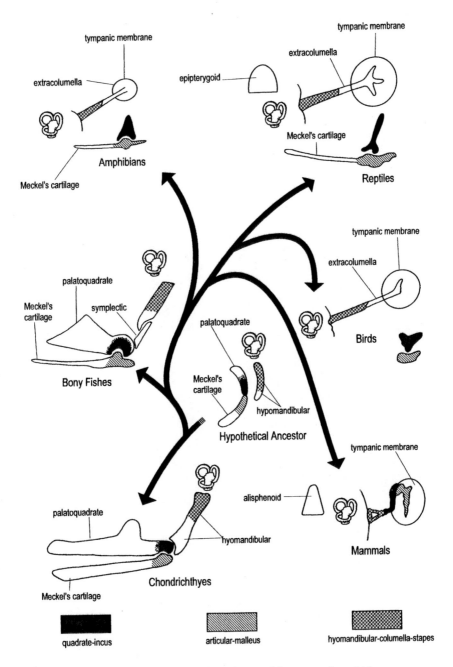

4.11 Diagram of the evolution of the three bones of the mammalian middle ear.
(Adapted from Webster and Webster 1974)

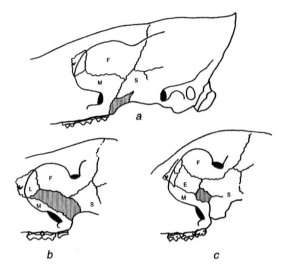

4.12 The relationship of the bones in the medial wall of the orbit: a) usual mammal condition (dog); b) lemurs; c) lorises, tarsiers, and anthropoids. F=frontal bone; S=sphenoid bone; E=ethmoid bone; M=maxillary bone. The palatine bone is lined. (Redrawn from Le Gros Clark 1971)

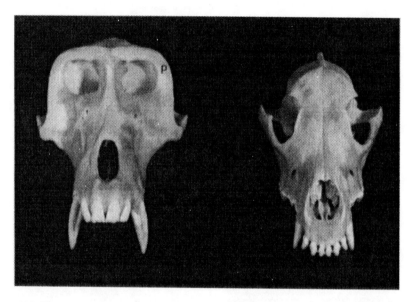

4.13 The postorbital bar (P) in a baboon. It is absent in the dog and many other mammals but is present in all extant primates. (Photo courtesy of Joseph R. Siebert)

ample, tree shrews, horses, and toothed whales. Prosimians possess only the bony bar (figs. 4.3 and 4.4). In higher primates, the back wall of the orbit has an additional bony partition separating it from the temporal fossa (figs. 4.5 to 4.7). This bony septum separates the orbital contents from the large temporal muscle occupying much of the temporal fossa and, in addition, affords attachment for its anterior fibers. Several adaptive explanations have been suggested for the origin of the anthropoid postorbital septum (Cartmill 1972, Cachel 1979, and Ross 1995). Cartmill suggests that it serves to improve visual acuity by insulating the orbital contents against muscle movements in the temporal fossa, Cachel believes that it serves as an additional attachment for the anterior temporal muscles, and Ross (1995, p. 275) maintains that the postorbital partition "is most likely" the result of anatomical conditions found in the fossil omomyids (omomyids are discussed in chap. 10). It is obvious that the case is not closed regarding the origin of the anthropoid postorbital septum.

When viewed from the lateral side, the skull shows several topographic features in the pterion (wing) region which are useful in distinguishing between platyrrhines and catarrhines. In platyrrhines the zygomatic bone is extensive and makes contact with the parietal bones (fig. 4.5), whereas in catarrhines this contact is excluded by the presence of the sphenoid and temporal bones articulating with the frontal bone (fig. 4.14), a condition that may only occasionally be found in New World monkeys. It is now clear that there is much variability in this respect, that is, that many combinations of bony contact are possible in these regions in all primates and that the differences and similarities are due, at least in part, to variations in rates (allometry) of early bone growth (Schultz 1963 and Mouri 1984).

When present, the mastoid process of the temporal bone is observable in the lateral view of some primate skulls (figs. 4.3 to 4.7). It is best developed in humans, but is occasionally present in the great apes, especially mature male gorillas (fig. 4.15). It contains the mastoid air cells, which are discussed below with the paranasal air sinuses. Also observe the different amounts of development of the sagittal crest in the two adult male gorillas shown in figure 4.15.

The Cranial Vault

The term "cranium" may refer only to the bones protecting the brain, in which case that portion of the skull is known as the cerebral cranium (fig. 4.14). The bones of the cranial vault are the frontal, the two parietals, the ethmoid, the two temporals, the occipital, and the sphenoid bone, united by fibrous sutures (joints between bones of the cranium and bones of the face). During infancy and early adolescence these sutures allow growth of the brain, face, and skull. As the animal becomes

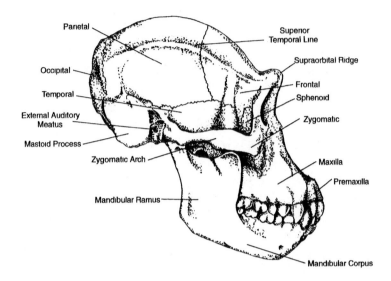

4.14 Side view of a male chimpanzee skull. (Redrawn from Swindler and Wood 1973)

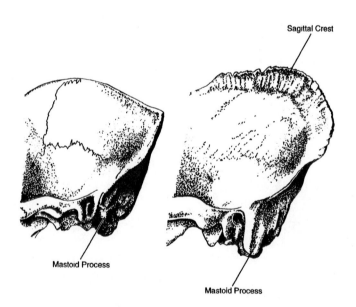

4.15 Relative sizes of the mastoid processes in two adult male gorillas. Also, note the difference in the development of the sagittal crests. (Redrawn from Schultz 1963)

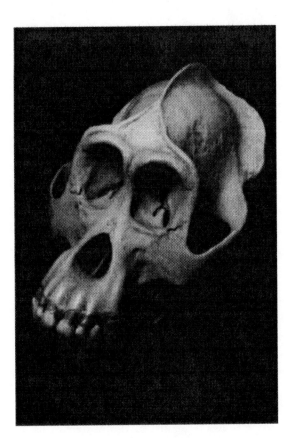

older the sutures lose the fibrous tissue between them, fuse, and become solid bone. The timing and pattern of suture closure are two of the biological criteria for estimating the approximate age of an animal's skull at death, especially in human forensic investigations.

In addition to sutures, the bones of the cranial vault display a rich collection of ridges, lines, crests, grooves, pits (fovea; pl., foveae), and holes (foramen; pl., foramina). Ridges, lines, and crests develop as a result of muscle attachments, for example, temporal lines for the temporal muscle and nuchal crests on the occipital bone for the muscles of the back of the neck. In many groups of primates (usually in males) a sagittal crest results when the temporal lines converge along the cranial vault for the attachment of the temporal muscles, as shown in the male orangutan in figure 4.16. Note also the large nuchal crest, which forms a veritable bony ledge in this male orangutan. Grooves, pits, and foramina result from blood vessels, nerves, and tendons passing on or through the bone as it develops. For example, blood vessels and nerves pass through the infraorbital foramen in the maxilla. The number of infra-orbital foramina indicates the number of vessels and nerves present when the bone developed around them, since bone develops after nerves and vessels embryologically (see figs. 4.6 and 4.7). The number of infraorbital foramina may vary greatly from one to several in members of the same species or may even vary from side to side in the same animal.

Basilar View of the Skull

The most obvious feature from the basal view of the skull (fig. 4.17) is the large hole through the occipital bone, the foramen magnum, through which the spinal cord passes to become the brain inside the cranium. As noted earlier, there has been a trend toward more forward and downward placement of the foramen magnum in anthropoids. An exception is in *Alouatta*, the howler monkey, where the foramen magnum opens more backward than downward. It has been suggested that this is a secondary change resulting from the repositioning of the head due to the extraordinarily large hyoid bone and mandible which are part of the unusual vocal apparatus of these primates (Le Gros Clark 1971).

On either side of the foramen magnum are the occipital condyles, which fit into a pair of depressions on the first cervical vertebra (the atlas). This important connection between skull and spine permits front-to-back motion of the skull. The joint between the atlas and the second vertebra (the axis) allows the head to glide from side to side. Thus, we nod yes at the first joint and shake no at the second joint. There are always two occipital condyles in mammals; reptiles have only one.

The bony external and middle ear regions display characteristics that are important in taxonomic and phylogenetic studies of primates. The middle ear region of all primates is enclosed within a bony chamber, the auditory bulla, which contains the three ear ossicles and either a simple or a specialized external auditory meatus. The auditory bulla develops from the petrous (rock) portion of the temporal bone in all living primates. In prosimians and most platyrrhines the bulla is inflated into a relatively large, elongated structure lying just lateral and anterior to the occipital condyles (fig. 4.18, and see figs. 4.3 and 4.4); while in catarrhines, the petrosal is not inflated as a bulla (fig. 4.17, and see figs. 4.5 to 4.7). The external auditory meatus is formed from the ectotym-panic bone (tympanic annulus) in primates. This feature may be ex-

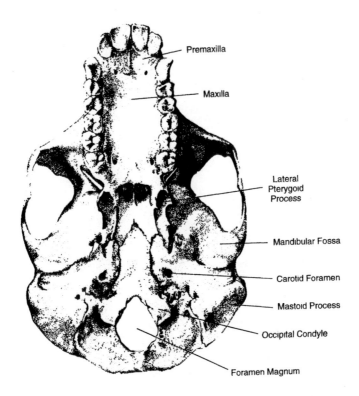

4.17 Basilar view of a chimpanzee skull. (Redrawn from Swindler and Wood 1973)

pressed as a ring lying within the bulla (lemurs) or attached to the wall of the bulla (as in lorises and New World monkeys). In tarsiers and catarrhines the tympanic annulus remains in contact with the temporal bone but is elongated into the external auditory meatus (fig. 4.19, and see figs. 4.2 to 4.7).

The type of external auditory meatus present in a specimen is one of the easiest and clearest ways of distinguishing between Old World and New World monkey skulls. Also, the bony relationships of the ear region of the skull and the presence of a postorbital bar are generally considered to be the most diagnostic bony features of the primate skull.

The mandibular fossa or glenoid cavity for the articulation of the lower jaw is part of the temporal bone and is seen in this view lying anterior to the bony external auditory meatus. The articular surface of the glenoid cavity is rather broad and flat in prosimians, New World monkeys, and Old World monkeys. It is more concave in hominoids, particularly in humans, where it becomes a deep, ovoid depression. There is a postglenoid process (a projection of the temporal bone) that

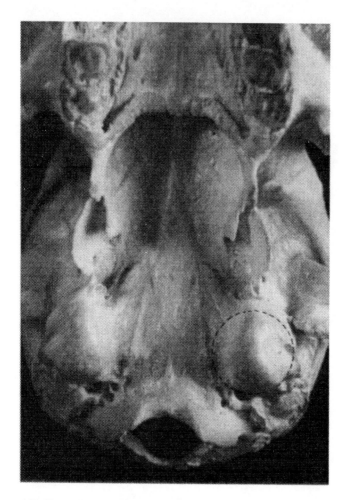

4.18 The auditory bulla in a prosimian, *Lemur catta*. Compare with figure 4.17. (Photo courtesy of Joseph R. Siebert)

forms the posterior wall of the fossa in all primates except *Daubentonia* (Ankel-Simons 1983). Ankel-Simons suggests that the absence of the postglenoid process is probably associated with the functions of the rodentlike jaw and dentition in the aye-aye.

Paranasal Sinuses

Several bones of the skull (the maxillary, frontal, ethmoid, sphenoid, vomer, temporal and palatine) may develop cavities or sinuses within the tissue between the two surfaces of the bone as they mature. This

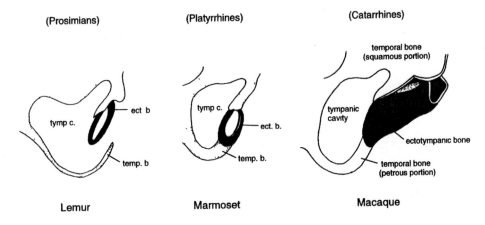

(Prosimians) (Platyrrhines) (Catarrhines)

temporal bone
(squamous portion)

ect b
tymp c.

tymp c.
ect. b.

tympanic
cavity

ectotympanic bone

temp. b
temp. b.

temporal bone
(petrous portion)

Lemur Marmoset Macaque

4.19 The structure of the bony external auditory meatus in primates. (Modified from Hershkovitz 1977)

process, known as pneumatization, is initiated by outgrowths of the mucous membrane of the nasal cavity and middle ear cavity. The mucous sacs wander into adjacent bones, slowly causing bone absorption, which may continue throughout life. The pneumatic cells of the middle ear normally invade the mastoid process of the temporal bone and, on occasion, may pneumatize contiguous parts of the temporal bone, as in the chimpanzee skull shown in figure 4.20. These bony sinuses are differentially present among primates and are all subject to considerable variation in form and size.

The paired maxillary sinuses would appear to be a primitive possession of the infraclass Eutheria (placental mammals) and are the largest of the sinuses in primates. They may be completely wanting in some catarrhines, for example, *Cercopithecus* and *Colobus* (Koppe and Hiroshi 1995). Other species may lack other sinuses. For example, there is no frontal sinus in *Pongo*, while the African apes have large frontal sinuses; the frontal sinus of *Gorilla*, in relative terms, is smaller than that of *Pan* (Blaney 1986). Cercopithecoid monkeys lack frontal sinuses.

You are aware of your sinuses, especially those in the frontal and maxillary bones, when you feel the pain that results when they become inflamed (sinusitis). The actual function of these bony cavities in primates remains somewhat elusive since they are probably multifunctional. They may, for example, influence vocalization, increase the area of the olfactory (smell) membrane, humidify and warm the air in the nasal cavity, lighten the skull, and add insulation and strength to the skull.

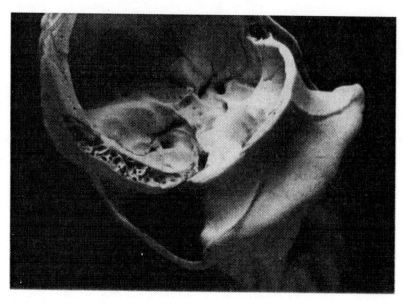

4.20 The cranial vault has been removed exposing the pneumatized portion of the temporal bone of an adult chimpanzee skull. (Photo courtesy of Joseph R. Siebert)

The Lower Jaw

The lower jaw, or mandible, contains the lower teeth. A triangular V-shaped mandible is the primitive mammalian condition and is found in most prosimians and many primitive platyrrhines, such as *Cebuella* and *Callithrix*. A U-shaped arcade is present in living catarrhines. The right and left halves of the mandible are fused early in development at the mandibular symphysis (cartilaginous joint) in the midline of the mandible in all higher primates, but remain loosely connected in most prosimians, permitting a certain amount of independent movement during chewing. It is generally thought that the fusion of the mandible in anthropoids is functionally related to the acquisition of vertical incisors.

On the inner side of the mandibular symphysis is the "simian shelf," an extremely variable bony ledge which extends horizontally behind the chin and forms a floor between the two halves of the mandible (fig. 4.21). This structure varies with age and sex and may be found in the lower jaws of all primate species except those of gibbons and humans (Hershkovitz 1977).

One other feature of the primate mandible warrants attention, the chin. By anatomical definition the chin refers to the symphysial region

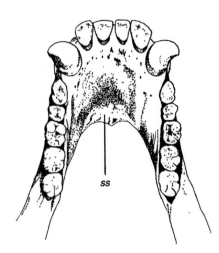

4.21 The "simian shelf" along the inner side of the front of the mandible of a chimpanzee. (Author's collection)

of the mandible, not just to what is called the mental (from mentum, meaning "chin" in Latin) protuberance of human anatomy (Hershkovitz 1970a). Thus, according to Hershkovitz, a chin is not unique to humans, since all primates have the symphysial region of the mandible. Only the human mental protuberance, which appeared late in human history, is unique in primates and has become a hallmark of our anatomy in recent times.

5 〜

Teeth,

Diet, and

Digestion

PITHECUS SATYRUS.

Teeth are composed of dense, hard materials, especially the enamel covering the tooth crown, and are the primary structures of mastication (chewing). Because dental structures are so hard, they are often fossilized when little else of an animal has survived. In many mammals, including primates, teeth have many additional functions, such as digging, holding and carrying, fighting, ripping, tearing, combing, cracking, and piercing. In a very true sense, teeth may be thought of as tools.

Teeth are of great importance to primatologists and paleontologists for several reasons. Teeth are variable among both fossil and living species, they change rather slowly through time, they are frequently specialized for particular functions, and the many morphological features of the teeth (e.g., cusp size, number, form, and groove patterns) are to some extent genetically determined. This means that related species will show certain similarities in their dental characteristics just as they do in other genetically controlled features. For example, molar morphology has contributed to taxonomic arrangements within species, genera, and families of both fossil and extant primates. Also, the anatomy of the tooth crown is broadly correlated with the different types of food eaten by primates. For these reasons, paleoanthropologist John Fleagle (1988, p. 15) has written, "Teeth, more than any other single part of the body, provide the basic information underlying much of our understanding of primate evolution." Thus, knowledge of odontology (from odont, Greek for tooth) is extremely important for understanding the evolution of the primates as well as the taxonomic diversification of contemporary forms.

TEETH

The teeth of primates, as in most mammals, are composed of four substances:

1. Enamel: a dense, hard cap covering and protecting the underlying dental tissues. It covers the crown of the tooth, which is the part of the tooth exposed above the gum. Enamel is the hardest substance in the body, varying from 5 to 8 on the Moh's scale of hardness of minerals, in which 1 represents the hardness of talc and diamond is 10.

2. Dentine (ivory): the main portion of the tooth. The dentine is covered by enamel on the crown and cementum on the root.

3. Cementum: harder than dentine but softer than enamel. This bone-like substance is arranged in layers around the tooth root.

4. Pulp: made up of soft tissues containing the blood vessels and nerves of the tooth. An important function of pulp is the formation of dentine.

A typical tooth (fig. 5.1) consists of a crown covered with enamel and roots of dentine enveloped in cementum. Dentine makes up the major portion of a tooth. A more or less well defined neck separates the crown from the roots. These three structures—crown, neck, and root—make it generally possible to distinguish the teeth of mammals from nonmammals (Peyer 1968).

Primates have a heterodont dentition, that is, the teeth are shaped differently from front to back. The incisors are cutting teeth, usually with a straight edge; the canines are piercing structures with a single, pointed

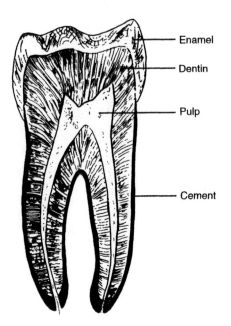

5.1 The components of a mammalian tooth.

Enamel

Dentin

Pulp

Cement

cusp; the premolars and molars are multicusped teeth for crushing and grinding. Some mammals (notably sea mammals) have a homodont dentition, that is, the teeth are of similar shape from front to back. The teeth in primates sit in and are surrounded by a bony socket. Blood vessels and nerves enter the tooth through the apical foramen (opening) at the tip of the root. The teeth of each side of the upper and lower jaws are generally mirror images of one another. The upper and lower cusps of the molars have been labeled with the zoological terminology used by scientists who study the dentitions of mammals (fig. 5.2).

Primates have two sets of teeth during their lives: deciduous (also known as milk, baby, and primary) and permanent. The deciduous teeth (incisors, canines, and molars) come in before the permanent incisors, canines, and premolars that replace them. The deciduous teeth are shed as they are replaced by their permanent successors. The permanent molars develop later as one set, that is, there are no deciduous teeth occupying the space in the jaws where the permanent molars develop and eventually erupt.

The number and class of teeth in each quadrant of the jaw can be written for both deciduous and permanent dentitions as a dental formula:

(1) 2i-1c-2m/2i-1c-2m = 20   (2) 2I-1C-2P-3M/2I-1C-2P-3M = 32

These dental formulas are the tooth number and class of tooth for (1) the deciduous (lower case letters) and (2) the permanent (capital

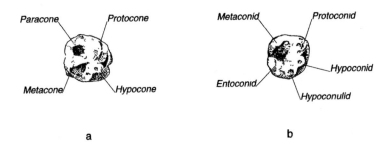

m

Paracone Protocone

Metacone Hypocone

a

Metaconid Protoconid

Hypoconid

Entoconid

Hypoconulid

b

5.2 Zoological terminology for the names of the cusps of mammalian molar teeth: a) upper right molar; b) lower right molar; m) refers to the mesial or front of the tooth.

letters). In this example, the formulas represent the deciduous and permanent dentitions for all catarrhine primates. The teeth to the left of the slash mark are maxillary teeth while those to the right are mandibular teeth. In table 5.1 (p. 108) the dental formulas are listed for several superfamilies and families of living primates. Note that the basic eutherian (placental) mammalian permanent dental formula was

$$3I\text{-}1C\text{-}4P\text{-}3M/3I\text{-}1C\text{-}4P\text{-}3M = 44$$

which is the number of permanent teeth possessed by most of these early mammals. Most modern mammals have modified the formula and reduced their full complement of teeth to fewer than 44. There are some exceptions such as the modern pig, which still has 44 permanent teeth.

The formulas presented in table 5.1 clearly indicate that several important dental reductions have occurred in primates during their evolutionary history. The most obvious tooth reduction in living primates has been the loss of the lateral incisor. It has been stated that having two incisors is a major diagnostic characteristic of the primate order (Le Gros Clark 1971), even though some of the basal groups of the Paleocene may have had three incisors (see chap. 10). Another reduction was the loss of premolars, an event that occurred from the front of the jaw back. P1 was lost before P2. The loss of a tooth does not alter the numbering of the remainder. All catarrhines, for example, have lost P1 and P2 and their remaining two premolars are phylogenetically identified as P3 and P4, not P1 and P2, although your dentist will refer to them as P1 and P2, or even call them bicuspid teeth.

The dentition of the modern Lemuridae has undergone a reduction in the premolar region from four to three while the Indriidae have lost two premolars. The Indriidae have the same number of premolars as do the Cercopithecidae and Hominoidea. Three premolars are present in all New World monkeys. Notice that the only molar reduction in living primates has occurred in the Callitrichidae, which have lost their third molars, except for *Callimico*. M3 is also occasionally missing in humans and gibbons. On the other hand, supernumerary M4s are sometimes found in orangutans.

Living tarsiers have only the lower central incisor, and it is implanted vertically. These vertically implanted lower incisors and canines may be an adaptation to their well-known habit of killing their prey by biting with their anterior teeth, the incisors and canines. Their diet, incidentally, consists almost exclusively of insects and small vertebrates (Richard 1985).

The most drastic dental reduction has occurred in Daubentonia. The aye-aye has lost an additional incisor, both canines, all lower premolars, and three upper premolars (fig. 5.3). The remaining enlarged incisors continue to grow (erupt) throughout life, reminiscent of the situation in rodents (see fig. 4.2). Incisors are used to gnaw bark and wood as the animal searches for the grubs and insects that constitute the major portion of its diet. At one time the aye-aye was believed to be a rodent, but it is now properly classified as a primate.

The evolutionary consequence of dental reduction is that once a tooth has been lost in a particular species it is lost forever. It does not reappear at some later time in the species's history. Such evolutionary phenomena illustrate a phenomenon known as the irreversibility of evolution. During phylogeny animals have returned functionally but not structurally to an ancestral condition; for example, the Ichthyosaur, a marine reptile of the Mesozoic era, evolved from a fish to a land tetrapod, then returned to the sea. In so doing, it did not lose its lungs but remained an air-breathing animal; nor did it re-evolve fins. Rather its fore- and hindlimbs became paddlelike structures that moved the creature through the water. There was a return in form and function but never a return to the original structure.

Because of genetic regulation, the number of teeth present in different taxa are important in primate taxonomy. The number of teeth in a species can help in identification. New World monkeys, for example, have three premolars in each quadrant of the jaw, while Old World monkeys have two. This is one of two easy ways to distinguish between the skulls of New and Old World monkeys. Look at the ear region. Is the external auditory meatus a bony tube (OWM) or a bony ring

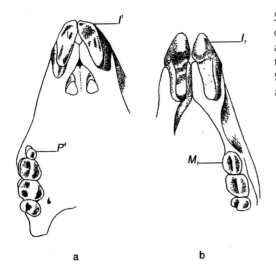

5.3 Occlusal view of the dentition of the aye-aye: a) upper teeth; b) lower teeth. (Redrawn from Swindler 1976, approximately × 2.0)

(NWM)? Count the number of premolars. Are there two (OWM) or three (NWM)?

Prosimii

Dental formulas and dental morphology are useful criteria for identifying the various groups of contemporary and fossil primates. Dental formulas are helpful in separating higher taxonomic categories, while the details of dental anatomy are frequently useful in sorting out individual species, and as we shall see in our discussion of fossil primates in chapter ten, minute differences in dental structures are frequently used in identifying fossil species.

All contemporary lemurs except *Lepilemur* have the dental formula listed in table 5.1. In this species, the maxillary incisors are absent in the permanent dentition, although they are present in the deciduous dentition. In most other forms the upper incisors are small and the central incisors are separated from each other, forming a space (diastema) between them. The lower incisors of all lemurs, indrids, and lorisids are procumbent (inclined forward), while in tarsiers the single lower central incisor is almost vertical (fig. 5.4). The incisors of the lemurs and lorises are joined laterally by the incisiform canines, forming what is known as a tooth comb. The function of the dental comb has been debated for years, but recent studies suggest that it is used differently among the species possessing it (Buettner-Janusch and Andrew 1962,

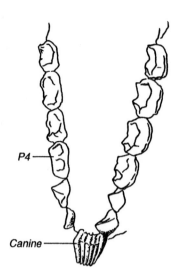

5.4 Occlusal view of the lower dentition of *Hapalemur griseus*, showing the lemur tooth comb consisting of the incisors and canines. Also, note the molarization of P4 which is characteristic of this species. (Redrawn from Swindler 1976, approximately × 2)

P4

Canine

and see below). One species may use it for self-grooming or social grooming, another species may use it to gouge and dig holes in trees to stimulate the flow of resin which can then be scraped up. In one example, the sifaka (*Propithecus verreauxi*), it has been observed that the tooth comb is used for both grooming and scraping (Richard 1985).

The canines have remained relatively stable (one canine per quadrant) in both the upper and lower jaws except in the forms where the lower canines have migrated mesially (or forward) to become incisiform, forming the lateral sides of the procumbent dental combs.

Contemporary prosimians display either two premolars (indriids) or three (lemurs, lorises, and tarsiers) except for the aye-aye, which has only the upper fourth premolar. The most anterior upper premolar (P2 or P3) is frequently caniniform (possesses a single pointed cusp), while P4 may be molarized (several cusps present as in a molar). Molarization of the upper P4 is most pronounced in lemurs and galagos. In the lower jaw premolars are different in size and morphology from P2 to P4, a condition known as heteromorphic premolars. In galagos and the gentle lemur, *Hapalemur griseus*, both upper and lower P4s are molarized to the point where they are nearly as large as the first permanent molars (fig. 5.4). The P4s are particularly large in the gentle lemur. This greatly increases the occlusal area in the region where major chewing takes place in these vegetarians that eat fruits, leaves, flowers, and stiff, fibrous reeds (Napier and Napier 1967).

The upper molars have either three cusps (tritubercular) or four (quadritubercular). The hypocone was the last cusp to appear phylogenetically and is the most variable cusp on the upper molars in all primates. It is generally the smallest cusp on the upper molars and shows a size reduction from M1 to M3. It is frequently quite small, particularly on M3, and is absent on all three upper molars in tarsiers. The lower molars generally have four cusps on M1 and M2, and occasionally a fifth cusp is present on M3.

The dentition of the aye-aye, mentioned a few times before, deserves special attention (see fig. 5.3). The dental formula for aye-ayes is quite unusual among primates (see table 5.1). There is only one premolar, the small, peg-shaped upper P4. The molars have four low cusps, which are worn smooth very soon after eruption, leaving a flat occlusal surface. Some investigators believe the upper and lower incisors have been lost and replaced by the continuously growing canines which have migrated forward to replace them. Others believe the front teeth are really the central incisors and that the lateral incisor and canine have been lost. The large, powerful incisors are used to strip bark from trees so that, when the larvae inside are located, the long, spindly third finger can be used to stir the larvae into a puree before this dinner is removed with the finger for eating (Richard 1985). Some twenty years ago Matt Cartmill (1972) suggested that the aye-aye occupies the ecological niche on Madagascar usually inhabited by the woodpecker. The specialized incisors and elongated third finger function as the woodpecker's beak and tongue.

Anthropoidea

NEW WORLD MONKEYS   The Callitrichidae are represented today in South America by the tamarins and marmosets. The former are essentially frugivores while the latter are gummivores, although both will supplement their diets with other foods. A dental feature is used for separating the two groups: the frugivorous tamarins have canines much longer than incisors (long-tusked), while the gummivorous marmosets have incisors as long as their canines (short-tusked) and use these lower incisors and canines to gouge the bark of trees and scrape up the flowing resin (Sussman and Kinzey 1984; see fig. 5.5).

The upper premolars have two cusps. The lower premolars grade from a caniniform P2 to two or three cusps on P3 and P4.

The maxillary molars generally have three cusps, although a small hypocone may be present in some species. The lower molars have four cusps; the hypoconulid is absent. Both upper and lower third molars are absent in the Callithricidae except for *Callimico*.

Table 5.1 Dental Formulae of the
Permanent Dentition of Living Primate Families

Prosimii
Lemuridae*
     2I-1C-3PM-3M / 2I-1C-3PM-3M
Indriidae
     2I-1C-2PM-3M / 1I-1C-2PM-3M
Daubentonidae
     1I- 0C-1PM-3M / 1I-0C-0PM-3M
Lorisidae
     2I-1C-3PM-3M / 2I-1C-3PM-3M
Tarsiidae
     2I-1C-3PM-3M / 1I-1C-3PM-3M
Anthropoidea
Callitrichidae+
     2I-1C-3PM-2M / 2I-1C-3PM-2M
Cebidae
     2I-1C-3PM-3M / 2I-1C-3PM-3M
Cercopithecidae
     2I-1C-2PM-3M / 2I-1C-2PM-3M
Hylobatidae
     2I-1C-2PM-3M / 2I-1C-2PM-3M
Pongidae
     2I-1C-2PM-3M / 2I-1C-2PM-3M
Hominidae
     2I-1C-2PM-3M / 2I-1C-2PM-3M

---

*Lepilemur
     0I-1C-3PM-3M / 2I-1C-3PM-3M
+Callimico
     2I-1C-3PM-3M / 2I-1C-3PM-3M

The other family of platyrrhines, the Cebidae, are mostly fruit eaters except for the howler monkey, *Alouatta*, which prefers leaves. The upper and lower incisors are variable in size and shape among the taxa, and in *Aotus*, the night monkey, the upper lateral incisor is quite caniniform.

The canines vary a great deal, but they are always longer and more pointed than adjacent teeth. There is some degree of sexual dimorphism in canine size in all species, and the differences are more pronounced in the larger forms.

The upper premolars generally have two cusps. The lower premolars tend to add cusps from P2 to P4—thus, P2 has a single cusp, P3 usually two cusps, and P4 may have three cusps.

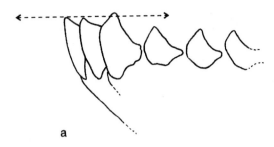

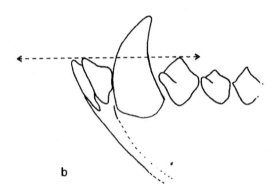

5.5 The lower canine-incisor relationship in the Callitrichidae: a) a short-tusked marmoset, *Cebuella pygmaea*; b) a long-tusked tamarin, *Leontideus rosalia*. (Redrawn from Swindler 1976)

The three maxillary molars are generally square and have four cusps. The hypocone is reduced from M1 to M3, where it is often absent. The lower molars also have four cusps except in *Ateles*, *Alouatta*, and *Brachyteles*, where the hypoconulid is added to M3. The most complicated molars in all cebids are in *Alouatta* and *Brachyteles*. Here a functionally integrated system of crests, styles, and basins is effectively adapted for crushing leaves (Swindler 1976).

OLD WORLD MONKEYS   As discussed earlier, the family Cercopithecidae is conveniently separated into leaf-eating and nonleaf-eating subfamilies. Members of the latter group are virtually omnivorous, preferring fruit, nuts, grass, seeds, and (for some species) even meat to leaves. Old World monkeys have reduced the number of premolars in each quadrant of the jaw to two, P3 and P4 (table 5.1).

The upper incisors are rather broad, especially I1, whereas I2 is narrow, often peg-shaped and pointed. The lower incisors are nondescript; I2 is slightly larger than I1. It is interesting to note that almost all Old World and New World monkeys have an edge-to-edge bite of the incisors (see fig. 4.8 of gelada baboon and fig. 5.7). An exception is in

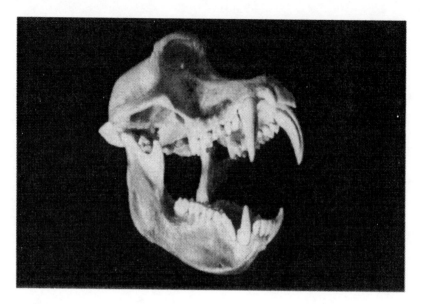

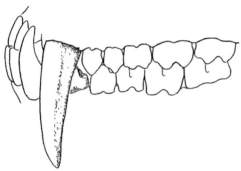

5.6 The large upper and lower canines of the male gelada baboon (*Theropithecus gelada*). Note also the elongated lower sectorial P3 and the deep groove on the mesial surface of the upper canine. (Photo courtesy of Nina G. Jablonski)

5.7 The upper canine and lower third premolar (C/P3) honing mechanism is present in Old World monkeys and anthropoid apes. It is not found in fossil or extant hominids. Note the edge-to-edge occlusion of the upper and lower incisors, which is present in most species of nonhuman primates. Leaf-eating monkeys, however, frequently have an underbite, that is, the lower incisors bite in front of the upper incisors.

leaf-eating monkeys which most commonly have an underbite, that is, their lower incisors bite in front of the upper incisors during mastication (see fig. 4.5, lateral view of howler monkey). Since the underbite is found only in leaf-eating monkeys it has been considered an adaptation for eating leaves (Zingeser 1970).

The upper canines are large and projecting teeth in both sexes. They are always larger in males, however, particularly in the more ground-dwelling forms. The lower canines are smaller than the uppers, but they too extend above the level of the other teeth. Canines are used by many primates for holding and preparing certain foods, but the large male canines of the more terrestrial species are used in agonistic (aggressive) behaviors as well as for threatening interactions between males (fig. 5.6). The large male canines offer good defense against predators, valuable since most terrestrial forms live in open country where the presence of predators is a constant threat.

The upper premolars are quite variable with respect to the number of cusps. P3 may have one or two cusps, while P4 often has as many as three. The lower premolars are heteromorphic: P3 has a single, usually elongated cusp, while P4 has two to four cusps. The elongated P3 is called sectorial (from *sector*, to cut) and describes the single cusp, compressed from side to side and forming a cutting edge that shears or hones against the lingual (tongue) side of the upper canine and is present in all cercopithecids. This C/P3 honing mechanism maintains a sharp edge to the back of the upper canine as well as an elongated cutting surface on P3 during chewing (fig. 5.7; also see the elongated lower P3 in gelada, fig. 5.6). P4 has two to four cusps, and in some species (e.g., *Papio*), it has become rather molariform.

A hallmark of all Old World monkey molars is an occlusal pattern found among primates only in these animals. The molars are bilophodont (fig. 5.8). Each molar has four cusps, two buccal and two lingual, connected by a transverse enamel crest (loph). The molars are constricted between the two sets of cusps, resulting in mesial and distal parts. The lower molars have an additional cusp, the hypoconulid, on M3 in all genera except *Cercopithecus* and *Erythrocebus*.

Anthropoid Apes

These primates are essentially vegetarians, preferring fruits and leaves to other types of food. Chimpanzees will supplement their diet with ants, termites, and an occasional baboon or colobus monkey.

In modern gibbons, the upper incisors are heteromorphic. The central incisor is broad with a concave lingual surface, while I2 is narrow and rather pointed. The lower ones are about equal in size. The canines are long, almost saberlike structures displaying little or no sexual dimorphism. This has been attributed, at least in part, to the gibbons' monogamous social structure. In addition, there is little male dominance within the group, that is, females are equally dominant in vocal-

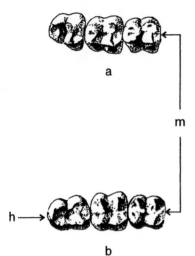

a

m

h

b

5.8 Bilophodont molars of an Old World monkey, *Cercocebus albigena,* the grey-cheeked mangaby of Africa: a) upper molars; b) lower molars; m) refers to the mesial or front of the tooth; h) refers to the hypoconulid.

ization and intra-group activities, both of which have been correlated to the lack of sexual dimorphism.

The upper premolars are bicuspid while the lower premolars are heterodont, P3 sectorial, P4 with several cusps.

The upper molars have four cusps in apes. The hypocone is the smallest cusp, as it is in most primates. The protocone is connected to the metacone by a crest, the crista obliqua, which is present in all hominoids. The hypoconulid is present on all lower molars. This results in the "Y5 occlusal pattern" that is characteristic of the lower molars of all extinct and living hominoids (fig. 5.9). The Y5 or *Dryopithecus* pattern is present when the metaconid contacts the hypoconid, and the buccal groove is mesial to the lingual groove when viewed from the lingual side of the tooth. This is an ancient pattern first found and described in the Miocene dryopithecines by Gregory (1916).

The pongids have broad upper central incisors with much narrower laterals. The I2s of orangutans are frequently quite pointed. The lower incisors are subequal in size, and their morphology is similar. The upper and lower canines are large, projecting teeth, especially the uppers. There is also greater sexual dimorphism between the upper canines. Incidentally, sexual dimorphism is more pronounced in gorillas and orangutans than in chimpanzees. There are many instances when it is not possible to separate male and female chimpanzees on the basis of canine size.

There are diastemata between the upper canine and upper lateral incisor and between the lower canines and P3. The former diastema re-

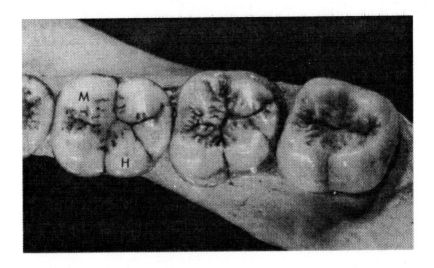

5.9 The Y5 occlusal pattern. This pattern is found in different frequencies on the lower molars of all fossil and extant hominoids. M=metaconid and H=hypoconid. In the Y5 pattern these two cusps are in contact, as shown on MI in this chimpanzee. The lower side of the tooth is the buccal side; the upper is the lingual.

ceives the lower canine, while the latter diastema admits the upper canine. These are present in the jaws of most primates but are quite variable and may not be present at all, as we mentioned in the last chapter.

The upper premolars are generally bicuspid while the lower premolars are heteromorphic; P3 is sectorial and hones against the lingual surface of the upper canine; P4 is bicuspid and frequently has three or four cusps.

The upper molars have four cusps, of which the hypocone is always the smallest. The crista obliqua is present. The lower molars consist of five cusps arranged in the Y5 pattern. There is some modification of the pattern from MI to M3 by the reduction of the hypoconulid and rearranging of the grooves into a plus (+) pattern (see the M3 in fig. 5.9). The Y5 pattern is most consistent on all three molars in gorillas and orangutans and shows more modification toward the plus patterns in chimpanzees, particularly, on M3 (Schuman and Brace 1955).

Hominidae

The teeth of humans are similar to those of anthropoid apes; consequently, only the major differences in the dentitions will be discussed here.

The most obvious and perhaps the most important disparity is in the size of the canines, particularly the upper canine. In humans the canine crowns are reduced in all dimensions and rarely extend beyond the crown tips of the adjacent teeth. The decreased size of the human canine is present in both sexes and is of little or no value in distinguishing between sexes.

The upper incisors are rather broad, with flat incisal borders, and are generally considered homodont, in contrast to the more heterodont incisors of pongids.

Both upper and lower premolars generally have two cusps, whereas the lower P3 in pongids is sectorial. The P3/C honing mechanism so characteristic of catarrhines has been modified in humans by the reduction of the upper canine and the change from a sectorial to a bicuspid lower third premolar.

The human upper molars usually have four cusps, as do the molars of pongids, but there is more reduction in cusp size from M1 to M3, especially the hypocone, resulting in more frequent three-cusped M3s than in pongids. Also, the cusps are generally less high and more rounded in humans. The lower molars tend to rearrange the (Y5) pattern by a change in cusp relations and initial reduction of the hypoconulid. This transformation of occlusal pattern results in a (+) pattern. With the complete loss of the hypoconulid, the cusp arrangement becomes a +4 pattern that is most common on M3 in most contemporary human populations. Incidentally, there is a feeling among some dental anthropologists that the +4 pattern is the evolutionary harbinger for the lower molars of humans and that both upper and lower M3s are on their way out.

Diet

There is an old German proverb that says "Man ist was er isst," which literally translated means that "one is what one eats." This is true for all animals since food contains the essential chemicals required for maintaining life: proteins, vitamins, carbohydrates, water, and minerals. In addition to the previous comments of the specific diet and dentition relationships, there are other issues to consider.

Primates commonly eat one type of food more often than another, although no primate species has specialized to the extent of the koala bear of Australia, which eats only eucalyptus leaves, or the giant panda of China, which subsists solely on bamboo. The aye-aye is probably the most specialized feeder among living primates, with its equally specialized dentition, but even the aye-aye will on occasion tear and gnaw coconuts and other fruits (Kavanagh 1984). Even the seed- and grass-eating gelada baboons of Ethiopia eat other types of food from time to time, as do leaf-eating monkeys (see table 8.1).

Modern primates are estimated to spend from 40 percent to 80 percent of their annual feeding time (Richard 1985) on one of six varieties of food (insects, saps and gums, fruit, leaves, seeds, and ground herbs) and are known respectively as insectivores, gummivores, frugivores, folivores, graminivores, and herb-eaters. The term "omnivore" is used to describe an animal that eats both animal and plant food. Humans are omnivorous. Many other primates eat both types of food, but they customarily prefer one to the other. Of course any classification is always somewhat arbitrary. Natural phenomena unfortunately do not always fit so conveniently within the boundaries we set. In other words, a primate listed as frugivorous may also include insects in its diet or eat different types of fruit during the course of a year.

The activity budgets of primates are closely related to their diets. Primates wake up hungry and usually begin to eat shortly after they get up. They generally spend a portion of this early activity time feeding on some high-energy food such as fruits or young leaves. The amount of time spent on feeding depends a great deal on the diet. For example, leaf-eating monkeys such as the colobines spend the majority of their time feeding and digesting leaves because leaves are low in energy and take a long time to digest. Macaques, on the other hand, have time for more varied activities because they feed on higher quality food sources. Since some leaf-eating monkeys, such as the New World howler monkeys, have not evolved specialized digestive tracts, they spend a large part of the day resting (63 to 79 percent for some species), presumably digesting their bulky food (Crockett and Eisenberg 1987). It is obvious that different species differ from one another in the allocation of time within this budget.

Despite such problems, primate dental morphology has been found to correspond pretty well with diet (Kay 1975, Kay and Highlander 1978, Swindler and Sirianni 1975, and Benefit 1990). For some last examples: the incisors of frugivores are wider and more robust than are those of folivores, an apparent association with how the front teeth are used in preparing fruit or leaves for eating. The bilophodont molars of Old World monkeys have higher cusps, increased shearing blades, and wider basins for crushing in Colobinae than in the Cercopithecinae. Interestingly, leaf-eating prosimians and howler monkeys also have higher cusps and longer shearing blades than the more frugivorous species.

When animals eat, different types of food wear against the enamel on the crown of the tooth causing different types of scratches. A relatively new analytical tool is dental microwear analysis by microscopic examination of the wear patterns on teeth (Teaford 1994). Dental microwear analysis (DMA) has proved useful in comparisons between frugivorous and less-frugivorous lemurs. The fruit-eaters have a higher frequency of enamel pitting than do the less-frugivorous lemurs (Teaford 1994). DMA investigations have also been useful in their application to paleobiological problems, and it is here that future DMA holds promise for helping to elucidate the life styles of fossil animals.

DIGESTION

Digestion is the process by which food, placed in the mouth and worked on by the teeth, is made absorbable by breaking it down into simpler chemical compounds. The digestive tract, or alimentary canal, is where digestion takes place and is composed of a rather complex system of structures in all primates. The mouth is the first segment of the digestive system, and it is here that digestion begins. The salivary glands produce saliva, which passes into the mouth, where it begins to soften and digest food. The former function is quite important since as saliva begins to dissolve and moisten food the sense of taste is aroused (chap. 6). Although digestion begins in the mouth, at this stage it is only concerned with the splitting of starches by the enzyme amylase, which is secreted by the salivary glands. There are three major salivary glands in primates: the parotid, submandibular, and sublingual. The salivary glands are larger and more specialized in the leaf-eaters, such as *Alouatta*, and the Colobinae. The cercopithecine cheek pouches (mentioned earlier) are temporary storage areas for food hastily obtained in the presence of competitors. They are similar to the crop found in birds, and, interestingly, both storage sacs contain predigestive enzymes. The enzyme amylase, for example, is found in the cheek pouches of cercopithecines (Jacobsen 1970).

From the mouth, food passes through the esophagus into the stomach, which varies little among primates. The stomach continues digestion by secreting pepsin, a protein-splitting enzyme, and its necessary companion, hydrochloric acid, which is indispensable for the action of peptic digestion. The stomach is essentially a simple, globular sac in most primates (fig. 5.10). The major exception is in the Colobinae, where the stomach is greatly enlarged and sacculated to accommodate the specialized diet of leaves (fig. 5.11). When the band of muscle shown passing across the surface of the stomach in figure 5.11 contracts, it helps to maintain the individual sacculations. Except for these monkeys, these muscle bands are normally present only along the colon. This extreme anatomical development is related to the specialized diet of leaves requiring a more complicated stomach (somewhat reminiscent of the four "stomachs" of ruminants), an elongated intestine, and a large cecum for digestion.

Although all Old World monkeys eat some leaves, it is the colobines that have specialized in leaf-eating. Leaves contain both proteins and carbohydrates (cellulose) and are available all through the year in the rain forest. Unfortunately mammals cannot digest cellulose without the help of microflora in the gut that digest the cellulose. Even in the presence of microflora the process takes a long time to break down the cellulose, so folivorous mammals, including primates, have adapted various strategies. Colobine monkeys have the high, sharp molar cusps for chewing leaves discussed previously, as well as sacculated stomachs and enlarged intestinal tracts where slow fermentation and absorption take place. Folivorous lemurs have a large cecum where the microflora digests most of the cellulose. In addition, *lepilemur* is coprophagous (feces-eating), requiring the great amounts of fibrous material in its diet to be passed through the alimentary canal a second time. This process is also common in rabbits.

The remainder of the gastrointestinal tract is divided into the small and large intestine, cecum, and rectum. In primates, the simplest arrangement is one in which the small intestine is suspended from the dorsal (back) abdominal wall by a double fold of tissue, known as a mesentery, from the lower end of the stomach to the rectum. This arrangement allows for many coils of small intestine to occupy the abdominal cavity. In most prosimians, and all simians, as the small intestine develops it undergoes several rotations so that the connection of the small and large intestine (the ileocolic junction) comes to lie in the lower right side of the abdominal cavity.

The large intestine of primates is more variable than the small intestine. From the ileocolic junction it forms a somewhat horseshoelike configuration around the coils of the small intestine as it passes upward,

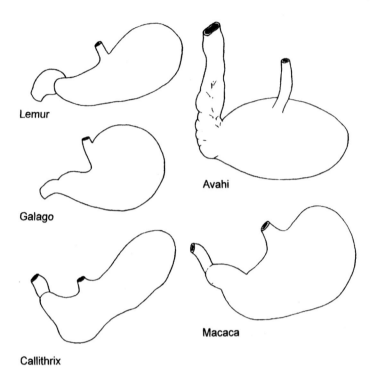

Lemur

Avahi

Galago

Callithrix

Macaca

5.10 The anterior (front) view of the stomachs of several nonhuman primates. The esophagus opens into the stomach on the right side and the duodenum exits to the left of the stomach. (Redrawn from Hill 1957)

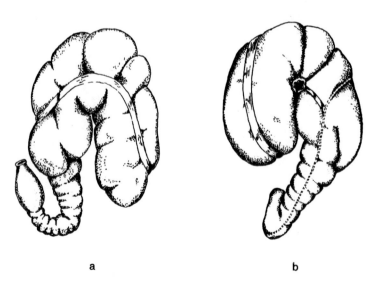

a

b

5.11 The sacculated stomach of the leaf-eating monkey *Colobus verus*: a) dorsoventral view; b) cranial view. (Redrawn from Hill 1957)

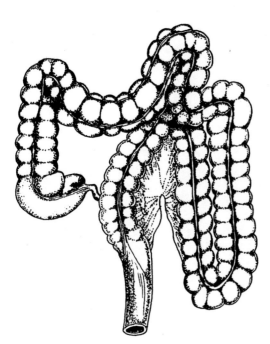

5.12 A view of the large
intestine of *Pongo pygmaeus.*
(Redrawn from Hill 1957)

transversally across, and downward along the left side of the dorsal abdominal wall to end as the rectum. Permanent sacculations occur along the length of the large intestine in Old World monkeys, anthropoid apes, and humans (fig. 5.12). They are not present in prosimians and occur only incipiently in marmosets, *Ateles*, *Brachyteles*, and *Lagothrix* among New World monkeys (Hill 1957). In many prosimians, the transverse colon forms an elongated U-shaped loop (*ansa coli*), which increases the length of the colon (fig. 5.13). This specialization is quite variable among prosimians; in some of the galagos (*G. senegalensis*), it is doubled upon itself (Hill 1957). The *ansa* is not present in any anthropoids.

The cecum—or lowest, most dependent part of the large intestine in the lower right side of the abdominal cavity—is present in all primates, though it varies greatly in size, position, and morphology. It may be sacculated in some prosimians, as in *Avahi* and pottos (fig. 5.13), whereas in *Lepilemur*, *Lemur*, and *Propithecus* it is rather capacious and coiled (Hill 1957). It is usually not sacculated in platyrrhines but generally possesses some sacculations in catarrhines. The position of the cecum in the lower right side of the abdominal cavity explains the position of the appendix, which is present among some primates, but is always present in anthropoid apes and humans. This troublesome prolongation of the cecum has outlived its usefulness and is often a source of trouble in humans

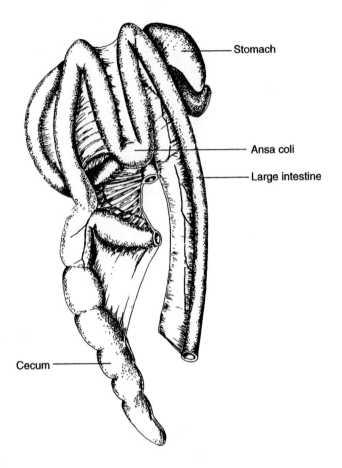

5.13 The *ansa coli* and the large sacculated cecum in *Arctocebus calabarensis*. (Redrawn from Hill 1957)

since it is a blind sac which can become inflamed and must then be removed surgically.

Digestion goes on in the small intestine as bile salts from the liver and pancreatic enzymes continue the digestive processes started in the stomach. After the material reaches the large intestine little absorption takes place and there is much bacterial action. Herbivores (leaf-eating monkeys) are dependent to a considerable extent upon the action of intestinal bacteria in the digestive process since the cellulose of plant cells is broken down only by cooking or bacterial action. In a recent study, Milton found that the colons of howler monkeys were much wider and longer than those of nonleaf-eating spider monkeys (see Radetsky 1995). The material would have to travel farther and would therefore remain longer in a colon that could support much more bulk allowing bacteria time to initiate fermentation and produce the energy-rich fatty acids so necessary for the howlers. The rather spacious colon and cecum of leaf-eating primates represent an adaptation necessary for the digestion of the animals' normal diet.

6

The Brain

and

Special

Senses

The brain or encephalon is the enlarged, greatly modified part of the central nervous system (CNS) enclosed in the cranial cavity. It is protected and supported by the bones of the skull, three fibrous membranes known as the meninges, and the cerebrospinal fluid. The primate brain is perhaps the single most distinctive character of the order. The brain matures early in its development, and consequently it is closely associated with reproduction. Thus, resources supplied by the mother to the fetus are critical during fetal life and play important roles in determining the eventual size of the brain. It is also well known that brain size in adults of a wide range of vertebrate groups including primates scales to adult body size with an allometric exponent value of about 0.75. It has been suggested by Martin (1990) that this scaling exponent of 0.75 is present because the basal metabolic rate of the mother acts as a constraint on fetal brain development. Basal metabolic rate (BMR) is a measure of the amount of oxygen consumed or the number of calories expended by an animal at rest over a prescribed period of time, for example, 24 hours. Thus, the maternal energy production during pregnancy is an important factor in the development of the primate brain.

The primate brain is distinguished from the brains of most groups of mammals by 1) its size relative to body size (1 : 20 in many monkeys and about 1 : 50 in humans; some sea mammals are exceptions); 2) expansion of the cortex (outer covering) of the brain, which is also called the neocortex (new cortex); 3) a richly grooved (convoluted) cortex; 4) a general increase in the visual system; and 5) certain evolutionary changes in the internal organization of the brain. A lengthy discussion of the evolution of the primate brain and its complex function in living primates is beyond the scope of this book. Here we will consider only the major trends that help demarcate the primate brain from that of other mammals.

The Gross Brain

The brain is subdivided into different parts (lobes) for convenience of description as well as for delineation of function. The paired cerebral hemispheres are the largest, most prominent portions of the primate brain, having undergone the greatest development during the evolution of the primates. The cerebrum is regionally separated into the frontal, parietal, temporal, and occipital lobes, which more or less correspond to the overlying bones of the skull (fig. 6.1). The superficial layer of the cerebral hemispheres is the cerebral cortex and consists of a rather thin layer of nerve cell bodies known as the gray matter. Below the cortex, the bulk of the tissue is the white matter composed mainly of nerve fibers.

Since the functions of the cerebral cortex involve memory, sensations, speech, vision, and motor activities, the cortex contains the association areas for regulating and processing this information (fig. 6.2 a and b). Thus, the amount of cortex is an important factor in the evolution of the brain since its development freed the organism from automatic and instinctive controls and made reflection based upon past experience possible. It is believed that the evolution of the primate neocortex was correlated initially with a change in the diets of early primates from insects to fruits. Remembering the location of fruit-bearing trees (visuospatial memory) would have been favored by natural selection in those primates in the process of shifting their diets (Allman 1982). This type of

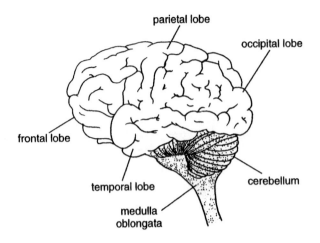

parietal lobe

occipital lobe

frontal lobe

temporal lobe

cerebellum

medulla
oblongata

6.1 The external surface of the human brain showing the major subdivisions (lobes) of the cerebrum. The cerebellum and medula oblongata are also shown.

feedback mechanism or loop was undoubtedly present in our prosimian ancestors. Occurring along with these changes, but perhaps slightly later in time, was the evolution of more complex systems of primate social organization that must have had important consequences for the development of the primate neocortex. Although the idea is controversial, it appears that the neocortex is constructed in orderly columns, or modules, and functions much like the microchips in computers (Eccles 1984). In humans, the cerebral cortex consists of approximately 2,000 square centimeters of surface area and contains 4 million modules, while a chimpanzee with 800 square centimeters of cerebral cortex would have 1.6 million modules (Jerison 1982). As the cortex expanded phylogenetically, it gradually overgrew an older part of the brain, the archipallium, which is the olfactory or smelling area. The size difference between these two portions of the brain is well marked in all primates, particularly in the apes and humans (fig. 6.3).

The major lobes of the cerebrum are defined by various fissures and sulci, which are frequently named. Only a few of these will be discussed here. The longitudinal fissure separates the right and left halves of the cerebrum. The lateral fissure (one of the oldest cerebral fissures) is a deep cleft between the frontal and temporal lobes on the lateral surface of the brain. The topographic isolation of the temporal lobe from the frontal lobe by the development of the lateral fissure occurred early in primate history. The separation of these two lobes by the lateral fissure and the even earlier development of the occipital lobe represent three

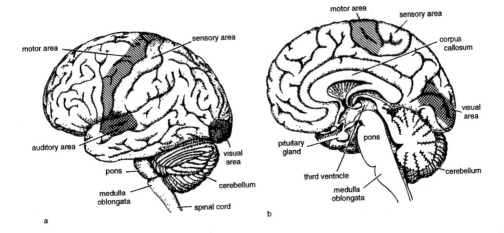

6.2 a) The external surface of the human brain showing the location of several of the functional areas. b) The human cerebral hemisphere seen from the medial side. The brain has been divided in the median plane.

characteristic features of the evolution of the primate brain. The temporal lobe has many responsibilities, including receiving and integrating much of the activity from all parts of the cortex. The expansion of the occipital lobe is due to the ever-increasing visual cortex of primates, which is located in the posterior part of the lobe.

The central sulcus is on the dorsolateral surface of the hemisphere that separates the frontal lobe from the parietal lobe. The portion of the frontal lobe in front of the central sulcus contains the primary motor cortex controlling voluntary movement, while the gyrus (a convolution of the brain) behind the central sulcus is the primary general sensory cortex receiving sensations from the body. The central sulcus is present in living anthropoids but absent in most prosimians, which instead have a coronolateral sulcus passing longitudinally through these somatic areas along the lateral side of the cerebrum (see fig. 6.3 and Radinsky 1975). According to Radinsky (1975) and Simons (1993), the central sulcus is present in *Aegyptopithecus*, an early anthropoid from the late Oligocene of Egypt, about 33 MYA.

A final cerebral sulcus deserves mention. This is the lunate, or so-called "simian" sulcus, which at one time was thought to be present only in the visual cortex of the occipital lobe of monkeys and apes (fig. 6.3). This sulcus, however, is also present in humans, but due to an enlargement of the parietal association region, it has been displaced to the medial side of the occipital lobe and is not always observable on the lateral surface of the cerebrum.

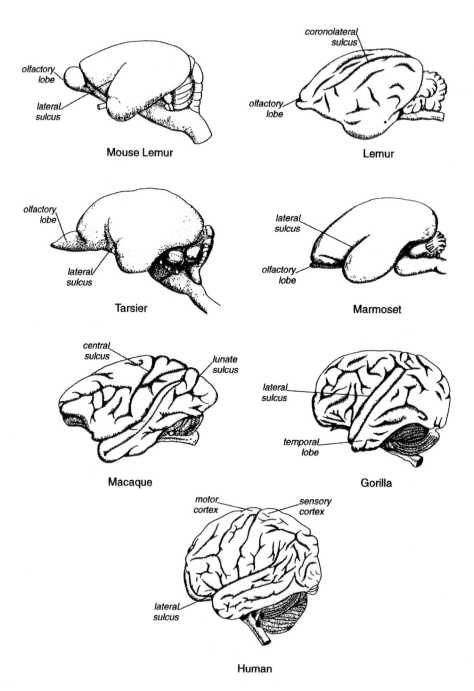

6.3 Comparisons of primate brains from the lateral side. Drawn approximately to the same size. (Redrawn from Le Gros Clark 1971 and Ankel-Simons 1983)

The cerebellum (fig. 6.1) develops from the upper portion of the brain stem to which it is attached. Its function is primarily concerned with the action of skeletal muscles, regulation of muscle tone, and maintenance of body equilibrium. Such function is important in the life of all mammals but becomes especially critical in primates, which have such varied locomotor habits. The internal structure of the cerebellum allows it to be in communication with areas throughout the body and, in turn, to relay information to higher centers in the cerebrum. It is interesting to note that the cerebellum of living primates, from prosimians to humans, is remarkably similar in its general morphology (Haines 1986). Yet as the primate cerebrum evolved, the cerebellum also underwent an expansion and elaboration resulting in numerous narrow folia (leaves) separated by narrow sulci (fig. 6.3). These folia and sulci are analogous to the gyri and sulci of the cerebrum. Thus, the human cerebellum has approximately 1,000 square centimeters of surface, but only about one-sixth of it is exposed (Ranson and Clark 1953).

The medulla oblongata (fig. 6.1) develops from the lower portion of the brain stem directly cranial to the spinal cord and contains the nerve cells for all cranial nerves except the first two, the olfactory and optic. The remaining cranial nerves control the activities of the face, tongue, pharynx, and larynx as well as some of the thoracic and abdominal viscera responsible for such basic physiological functions as respiration, temperature regulation, and heart rate.

Increase in brain size (encephalization) is perhaps the most obvious of the many anatomical trends that separate primates from other mammals. Primates as a group have the largest brains of all mammals. It is true that nonhuman primates do have large brains compared to many other land mammals, but when brain size/body size ratios are considered, it is found that some sea mammals (e.g., toothed whales) have a higher index than all primates exclusive of humans (Stephan 1972). Thus, "only man has encephalization, which exceeds that of all animals. He is the only Primate with an outstanding brain size" (Stephan 1972, p. 174). Indeed, Radinsky (1975) supported this claim when he pointed out that the human brain is some 3 to 3.5 times larger than might be expected in a higher primate with a human body weight. As we shall see in chapter eight, building larger brains is reproductively expensive and requires longer gestation periods with smaller litters (Pagel and Harvey 1988). It seems that one key to the evolution of larger brains "demands that a species inhabits a stable environment and experiences low mortality" (Lewin 1988, p. 514).

Another important factor when considering the evolution of the brain is the internal architecture or cytoarchitectonics, that is, studying quantitative differences in the internal organization of the brain and relating them to functional adaptations and possible phylogenetic relationships. As pointed out by Holloway (1968), some of the major differences among the brains of living primates probably involve neural reorganization, that is, shifts in quantitative components within the brain resulting in functional changes among the species. It should be remembered also that all parts of the brain did not evolve at the same rate. Differential growth, both phylogenetically and ontogenetically, is a hallmark of biological change. The primate brain is a good example of such mosaic evolution (Armstrong 1982).

Another difference between the brains of most mammals and those of primates, although there are differences among primates in this characteristic as well, is the degree of fissuration of the cerebral hemispheres (fig. 6.3). In more primitive mammals the surface of the cortex is smooth, whereas among most primates, fissuration results in convolutions, sulci, and fissures (but see Callithrix [marmoset] in fig. 6.3). As discussed earlier, fissuration of the surface of the cerebrum increases the surface area many times and is generally associated with an increase in the size of the cortex.

The brains of living primates present a wide range of anatomical variation, as seen in the lateral view of figure 6.3. Among strepsirhines, the most primitive brain configuration is that of Microcebus, the mouse lemur. The olfactory bulbs are reduced, as in other primates, but in the mouse lemur they are relatively larger than in other strepsirhines and are visible beneath the frontal lobes of the cerebrum (Le Gros Clark 1971). The cerebral hemispheres are smooth, lacking any fissuration except for the lateral fissure separating the temporal and frontal lobes. Among the larger lemurs the cerebral cortex is convoluted, and the olfactory bulbs are relatively more reduced but still visible lying under the frontal lobe. The sulci are directed somewhat longitudinally, and the central sulcus is no more than an irregular depression on the cortex (Le Gros Clark 1971). There are exceptions however in the position and degree of development of the central sulcus among the Lorisidae. In Perodicticus and Arctocebus this sulcus runs vertically between the motor and sensory regions as it does in most anthropoids (Haines et al. 1974 and Radinsky 1975). Among lorisids, the galagos generally appear to have fewer cerebral convolutions and sulci than other members of the family, but in his excellent study of the brains of lorisids, Haines (1974) found that the lesser galago has a more complex sulcal pattern than previously believed. The aye-aye, *Daubentonia*, possesses a brain that goes along with the rest of its anatomy in that it has primitive primate characteristics as well as features similar to those in nonprimate mammals. Indeed, Le Gros Clark (1971, p. 247) claimed that the sulci pattern in some specimens has "a curious resemblance to the sulci of a cat's brain."

One is immediately impressed with the virtual lack of fissuration in the brain of tarsiers (fig. 6.3). Even the lateral sulcus is absent, according to Le Gros Clark (1971), although Hershkovitz (1970a) claims that it is present and better defined in tarsiers than in the mouse lemur. It is present in the brains of all other living primates. The depressed regions visible along the anterolateral surfaces of the cerebrum are due to the large orbits that house the very large eyes of tarsiers (fig. 6.3 and see fig. 4.8). Parallel with this development is the great expansion of the visual cortex in the occipital lobe. The visuosensory area forms almost half of the cortex. At the same time, the olfactory complex is more reduced than it is in strepsirhines. The brain of tarsiers represents a complex of primitive and advanced features that result in an interesting evolutionary mosaic.

The cerebral fissural patterns in platyrrhines have been studied extensively by Hershkovitz (1970b and 1977). The most primitive cerebral fissural configuration was in *Cebuella*, the pygmy marmoset, which displays a smooth surface except for the presence of four sulci of which the lateral sulcus is the deepest and most prominent. The brain increases in size and cortical complexity in the other marmosets, whose brains may have as many as eight cortical fissures. There is much variation in cortical patterns among the cebids as well as continued reduction in the olfactory area and increase in the visual cortex. The only nocturnal monkey, *Aotus*, as might be expected, has a particularly well developed visual area. Among cebids, the most fissurated brains are found in the brachiating prehensile-tailed spider monkeys, *Ateles* and *Brachyteles*. It is interesting to note that *Alouatta*, the howler monkey, which is also a brachiating prehensile-tailed monkey, has a comparatively small brain with fewer than 20 cortical fissures (Hershkovitz 1970b). It is difficult to explain how the brain of such an active monkey has remained so smooth and relatively small.

The brain continues to enlarge and fissure patterns become more complicated in the Old World monkeys, pongids, and humans (fig. 6.3). The basic convolutional pattern of Old World monkeys is similar to that of apes and humans, except that, in the latter, secondary sulci develop that tend to obscure the primary sulci. The most complicated sulci and fissure pattern is found in humans and is related to their larger brain relative to body size ratio (ca. 1:50). There is, however, a great deal of variation among individuals in cortical fissuration. Cranial capacity also varies in contemporary humans and ranges from around 900 to about 2,000 cubic centimeters (average, 1,430 cc). Contrary to popular belief, there does not appear to be any direct relationship between cranial capacity and intelligence. For example, the great French novelist and wit Anatole France had a brain weight of only 1,017 grams, whereas the brain of the equally famous French paleontologist Georges Cuvier weighed in at 1,830 grams (Gould 1981). Among apes, the gorilla has the largest cranial capacity, with an average of 535 cubic centimeters, and is usually considered to have the convolutional pattern most similar to humans (fig. 6.3). See table 8.1 for the brain weight/body weight comparisons of several contemporary primates.

In summary, the evolution of the primate brain may be briefly characterized as follows: there has been a gradual increase in brain size relative to body size; the neocortex has enlarged more than would be expected from the increase in brain size; the olfactory bulbs and olfactory system have not become elaborated; the visual cortex has expanded; there has been an increase in the complexity of cortical fissuration relative to body size; an increase in gross brain size has resulted in major qualitative differences in behavior due to the geometric increase in the number of synaptic connections in the association cortex and between the cortex and the limbic system; and, finally, the differences between human and ape brains appear to be only quantitative.

SPECIAL SENSES

The special senses—vision, taste, smell, hearing, and tactile sensation—may all be said to be extensions of the brain since the special sense organs have neural centers located in the brain. Primates have elaborated and enhanced the discriminating abilities of some of these senses (vision) while reducing the sensitivities of others (smell). It is through these special senses that animals are connected with the physical world.

Vision

Primates are animals of great visual acuity who depend on this ability during their daily or nightly activities. Primates evolved in the trees, and most of them still spend a great deal of their time in trees today. Even those who forage on the ground—baboons, for example—will normally return to the trees when danger is near or for sleeping. To a greater degree than any other mammal, primates depend more on vision than on hearing or smelling. Consequently, the visual system in primates is an elaborate neural mechanism for acute vision that includes binocular vision in most (certainly in anthropoids and probably in diurnal prosimians to some extent) and color vision among the anthropoids and, again, probably in some prosimians.

The visual system is quite complicated in primates and only a few major features will be considered here. The eyeball is composed of several specialized layers, of which the innermost one, the retina, is the continuation of the optic nerve. The retina is made up of two types of cells, called rods and cones, which receive light stimulation from the outside. Rods are present in nocturnal animals and are sensitive to black and white stimulation, while cones are used for fine discrimination and spatial relationships and also possess the photopigments necessary for color vision. Cones are more common in diurnal forms. In primates a duplex retina (having both rods and cones) is found in all diurnal species; nocturnal prosimians possess only rods. The night monkey, *Aotus*, has mostly rods with a few cones. It is the only higher primate having a retina with almost all rods and only a single photopigment in its cones; consequently, it lacks color vision (Jacobs et al. 1993b).

There is some evidence that color vision predated the earliest mammals, and color vision is much more widely distributed among living mammals than originally thought. Indeed, Friedman (1967), after finding it in the North American opossum (*Didelphis*), claimed it to be a primitive mammalian trait. There is little information regarding the evolution of color vision among primates, but what there is suggests that the basis for trichromatic (cones possessing three types of photopigments) color vision occurred perhaps 30 MYA (Yokoyama and Yokoyama 1989). This was after the separation of the New and Old World monkeys and before the Old World monkeys and hominoids split (Jacobs 1995). It appears that color vision has been a part of the primate visual repertoire for a very long time.

Color vision is present in all diurnal primates. In Old World monkeys, apes, and humans, color vision is similar and known as trichromatic. Among New World monkeys the situation is more complex in that both trichromatic and dichromatic retinas are present (Jacobs 1995). Further, it is interesting to note that among New World monkeys the males are dichromatic while the females are both dichromatic and trichromatic (Jacobs et al. 1993a). Prosimians have not been widely studied for color vision, but in a recent investigation of the ring-tail lemur (*Lemur catta*) and the brown lemur (*Eulemur fulvus*) Jacobs and Deegan (1993) found that these lemurs, like most mammals, have the potential for dichromatic color vision. What this means with respect to color vision is that a trichromatic primate can make more subtle color distinctions than a dichromatic primate.

Napier and Napier (1985) state that color vision is not present in any nocturnal primate. Recently this has been supported by the work of Jacobs and colleagues (Jacobs et al. 1993a, Deegan and Jacobs 1994). In an excellent summary article on primate color vision Jacobs wrote, "It begins to appear that among primates one price of strict nocturnality is loss of color vision" (1995, p. 204).

Two other features on the inner surface of the central area of the retina are the macula lutea and the fovea centralis. The macula lutea (yellow spot) is a sort of filter composed only of cones, and the fovea is a small indentation of closely packed cones in the middle of the macula where visual acuity is greatest. These two structures are associated with diurnal animals (but also present in the nocturnal tarsier and the night monkey, *Aotus*) and could indicate that the ancestors of these primates were diurnal. According to Walls (1942), the fovea is also found in bony fish, some snakes, and birds.

A structure present in all nocturnal primates (except tarsiers and the night monkey) is the tapetum lucidum, a membrane lying on the outer surface of the retina which increases the animal's ability to see at night by reflecting light back through the retina. The tapetum also reflects light and is responsible for the eyes lighting up at night when a light is directed on them, for example, in dogs and cats. The eye is usually larger in nocturnal primates, and occasionally quite large, as in tarsiers where the eyes are almost 4.5 percent of their body weight (Schultz 1972) resulting in orbits that are larger than their brains (see fig. 4.8a). See figure 6.3 for the impression that the orbits leave on the sides of the tarsier brain. The slow potto and *Nycticebus*, however, have small eyes for their body size, especially the former.

Binocular or stereoscopic vision has been an important acquisition among primates because it allows judging distances between trees before leaping and permits close examination of objects held in their mobile hands. Binocular vision requires two structural modifications from monocular vision, a cranial one and a neural one. The cranial change involves rotation of the bony orbits from a lateral position to a more frontal orientation, which commenced in the Eocene primates and achieved maximum frontality in humans. This reorientation of the or-

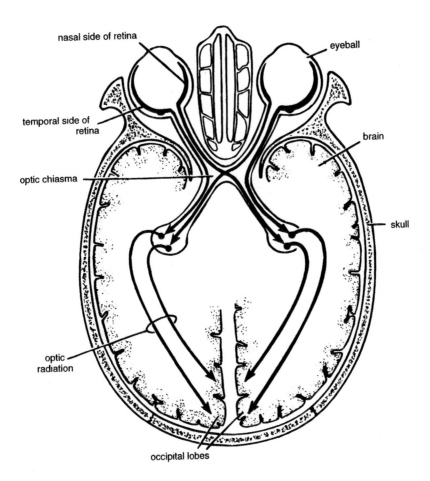

nasal side of retina

eyeball

temporal side of
retina

brain

optic chiasma

skull

optic
radiation

occipital lobes

6.4 Diagram of the neural basis of binocular vision of primates showing the optic chiasma and optic projections to the visual cortex in the occipital lobes. (Modified from Napier and Napier 1985)

bits to the front of the skull allowed the visual fields of the two eyes to overlap. The neural modification occurred along with the orbital change and resulted in the reorganization of the fibers of the optic nerve. In lower mammals, the fibers from one optic nerve cross at the optic chiasma to the opposite cerebral hemisphere (visual cortex); that is, fibers from the left eye cross to the right visual cortex, a condition known as decussation. With the development of binocular vision the fibers of the optic nerve begin to pass to the hemisphere of the same side. There are fewer and fewer fibers decussating, so there is now visual field overlap of corresponding areas of the two eyes on the same visual cortex (fig. 6.4). The superimposition of two images necessary for binocular vision is the function of the brain. Diurnal lemurs have binocular vision, as do humans and the other higher primates, but humans and the anthropoid apes appear to have fewer uncrossed fibers than the other primates, by about 40 percent.

Smell

Compared to most other mammals, primates have a reduced olfactory (smell) system. The olfactory system has not become elaborated in the course of primate evolution but has remained about the same over time. There is little doubt that olfaction is important for primates in certain environmental situations. The ability to smell is still present in primates and certainly plays a role in their social lives, particularly in prosimians and many New World monkeys. The point seems to be that it is less important relative to vision.

The nose in all prosimians except *Tarsius* consists of a moist, naked, rhinarium with a median groove or philtrum, as in mammals in general. The upper lip is also attached to the gum of the upper jaw along the philtrum and is immobile. This arrangement is quite different from that of *Tarsius* and all other primates, where there is no moist rhinarium and the upper lip is not attached to the upper jaw. The philtrum, if present, is only a shallow groove passing from the nasal septum to the bottom of the upper lip.

The type of nose a primate possesses is used in a system of classification proposed many years ago and still preferred by some primatologists (see chap. 2). In this classification the order is separated into two suborders, the Strepsirhini and Haplorhini. The former have a moist, hairless rhinarium and a median cleft and include all prosimians except *Tarsius*. The latter possess a dry nasal area and a mobile upper lip without an attached median cleft (see fig. 2.4). This suborder includes tarsiers, New and Old World monkeys, apes, and humans. The anatomy of the nose is also used to separate New World monkeys from Old World monkeys and apes. The former are the Platyrrhini, with nostrils pointing sideways; the latter are the Catarrhini, with downward directed nostrils (see fig. 2.4).

Smelling is a function of the olfactory nerve, which picks up odors from the lining of the upper regions of the nasal cavity and transmits them via several brain centers to the olfactory region of the brain. The nasal cavity contains several bony, scroll-like structures (conchae) projecting from the side walls into the cavity. These are covered with a mucous membrane containing the olfactory cells. The number and complexity of the conchae, and thus the size of the nasal cavity, is correlated with the ability to smell. In primates there has been a general reduction in the number of conchae, along with the olfactory bulb. An exception is the aye-aye, which has retained several more conchae than other primates and has a large olfactory bulb in the brain, indicating its greater dependence on the sense of smell. The anthropoids have all reduced the number of conchae, and their olfactory mucosa is limited to the upper recesses of the nasal cavity, a condition most apparent in the great apes and humans.

Prosimians have special scent glands located on their wrists, along their forearms, and in their armpits. These glands secrete pheromones (odors) that are used for scent-marking by rubbing them on branches, trees, or the ground. The scents are used to mark territories, attract mates, and communicate among the various groups. Scent-marking a territory can reduce the likelihood of actual combat when different groups are moving through the same areas. Pheromones can also herald information pertaining to a female's reproductive cycle. New World monkeys also have scent glands which are used to mark territories and convey information about sex and age. Old World monkeys and apes use visual communication more than scent, although odors undoubtedly play some role in their social behavior as they certainly do in human social behavior.

Taste

Taste buds in primates are located on the surface of the tongue, soft palate, and epiglottis. There are four basic taste sensations: sweet, sour, bitter, and salty. In humans, salt and sweet are picked up on the tip of the tongue, bitter at the base of the tongue, and sour along the borders. In order to taste any substance, it must be in solution. The function of the salivary glands mentioned earlier is to produce saliva which helps to dissolve food. It has been estimated that humans have about 3,000 taste buds.

According to Glaser (1972), there are no differences among primates in their ability to taste the four basic sensations. Of course, there are many degrees of sensations received from these four basic stimuli which are recognized by the taste receptors or taste buds. Further studies (Glaser et al. 1992), using sweet aspartame, tested a wide range of primates for their ability to taste this substance and found that only Old

World monkeys, apes, and humans could taste the sweet aspartame. This and earlier studies by Glaser and colleagues have led them to postulate the presence of different sweet taste receptors in the different species of primates.

Hearing

In primates, as in most mammals, there are three parts to the hearing apparatus: the external ear, the middle ear, and the inner ear.

The external ear of most primates is relatively small, varying in size and shape among the various groups. The basic anatomy, however, remains more or less similar (fig. 6.5). Occasionally the human ear has a small elevation arising from the inner edge of the upper rim known as Darwin's tubercle. There is little doubt that the size of the outer ear is related in some degree to sound reception, but there is also evidence that in some mammals, rabbits and elephants for example, the ears are temperature regulators. To what extent the external ears of primates function as temperature controls is not known, but it was pointed out several years ago by Le Gros Clark (1971, p. 281) that it seems probable that the difference in size between the small ears of the slow-moving lorises and the large ears of the fast-moving galagos "is related to different requirements for temperature regulation."

The external ears of primates possess several small intrinsic muscles which can move the various parts of the ear. The ears of some nocturnal prosimians (galagos) have ribs and folds, and the ears can be folded down, elevated, and drawn back (fig. 6.5). The external ears of all primates are mobile to some extent; even some humans can wiggle their ears. Tarsiers have quite large ears that are undoubtedly useful for picking up sound waves as they scurry through the brush at night looking for insects, frogs, and lizards. They also use their large ears in complex communications between individuals that consist of sophisticated, high-frequency "chirps" (Jablonski, personal communication).

The primate external bony portion of the auditory system was discussed earlier with the skull. Inside, the tympanic membrane or eardrum receives air waves from the outside environment and transmits them via the bony ear ossicles, malleus, incus, and stapes to the inner ear. The ear ossicles grow and develop to adult size by birth, completely within the middle ear cavity. Their shape and size, particularly of the malleus, are useful in primate taxonomy. The ear ossicles of pongids, for example, show more similarity with those of humans than with

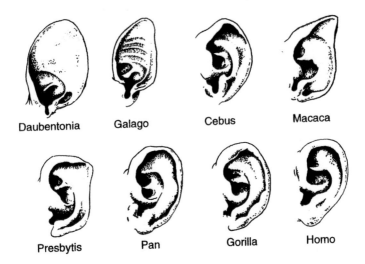

Daubentonia   Galago   Cebus   Macaca

Presbytis   Pan   Gorilla   Homo

6.5 The external ears of eight primates reduced to the same height. (Redrawn from Schultz 1963)

those of monkeys, while the ossicles of tarsiers are more like those of lemurs.

The inner ear contains three distinct sensory systems: the three semi-circular canals for detecting rotational movements of the head; the saccule and utricle for determining the position of gravity; and the principal organ of hearing, the organ of Corti. In general primates have good hearing, but there are some differences in the ability to hear certain frequencies. Chimpanzees and humans cannot hear the high frequencies that bush babies and marmosets can; humans, on the other hand, can hear quite low frequencies (Ankel-Simons 1983). An interesting proposal by Schultz (1972) suggests that it is likely that the long, loud sounds of siamangs and orangutans with their specialized throat pouches, as well as those produced by howler monkeys with their large laryngeal systems, may contain modulations that cannot be heard by other species.

Tactile Sense

The skin of primates is well supplied with sensory receptors to receive tactile pressure, pain, and heat stimuli. The vibrissae or sinus hairs (such as the bristly hairs around the mouth of cats) are sensitive to slight contact and are usually somewhat longer and thicker than other hairs (fig. 6.6). They are particularly well developed in prosimians, reduced

6.6 The position of vibrissae on a prosimian. These tactile hairs are localized into the supraorbital, upper lip, chin, cheek, and wrist areas. (Modified from Hill 1972)

in tarsiers, and usually only present in New World monkeys above the eyes and along the upper lip. When present in Old World monkeys they are limited to the face. They are absent in apes and humans. It is interesting to note, however, that in the young fetuses of some Old World monkeys long vibrissae have been found near the wrist; carpal vibrissae are present in human fetuses, but never present in adults. Thus, although tactile hairs are absent in adults of these species, they are present during their early development, "bearing witness to the original presence in human ancestry of the carpal vibrissae which characterize the lower primates" (Le Gros Clark 1971, p. 285).

Touch pads are present on the palms and soles of the hands and feet of all primates. The primitive arrangement of these pads is in three groups (proximal, intermediate, and distal), but they display various degrees of fusion among the different species, particularly in the higher primates (see fig. 2.5). They are well supplied with sensory nerve fibers containing pressure receptors so necessary in the grasping hands and feet of primates.

The skin covering these pads in primates forms the ridges and grooves that are the fingerprints (dermatoglyphics), which, as we know, are used for individual identification in humans. In addition to the many sensory nerve fibers present in the pads, there are numerous large sweat glands. This is critically important for all primates, because they enhance the function of grasping and holding objects. In some prosimi-

6.7 A six-month-old howler monkey (*Alouatta seniculus*) infant rides on its mother's back as she crosses from one shrub to another. The mother is aided by her prehensile tail. (Photo courtesy of Carolyn M. Crockett)

ans, most dramatically in tarsiers, the tips of the fingers and toes are expanded into round, flattened suction discs, which increase the surface area for more contact and better stability as they leap from branch to branch. Dermal ridges are also present on the distal half of the under surface of the prehensile tails of New World monkeys (woolly, howler, spider, and woolly spider). These greatly increase the tactile and grasping abilities of what we should probably think of as "fifth hands" (fig. 6.7).

7

The Skeleton

and

Locomotion

M. Seniculus Red Howler

The mammalian body can be divided into halves by means of three planes which are arranged with reference to length, breadth, and thickness. The three planes are very useful as landmarks in describing the various parts of the body. The saggital plane divides the body into right and left halves; the transverse plane divides the body into anterior and posterior halves; the frontal plane divides the body into dorsal and ventral halves (fig. 7.1). It is obvious that a quadrupedal animal moves forward with the ventral body-half underneath and its anterior parts in front, whereas humans move forward with the ventral body-half in front since they are tipped up on one end.

All mammals share a common basic skeletal structure that is inherited from a primitive, quadrupedal, probably arboreal mammal having four limbs of approximately equal length and carrying the body horizontally. Mammals have evolved a number of skeletal specializations based upon this structure. The skeleton of contemporary primates shows

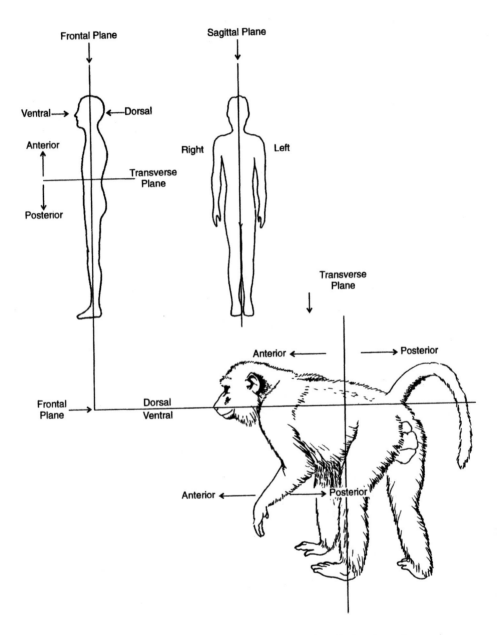

7.1 The planes of symmetry in bilaterally symmetrical animals, with the terms of position. The sagittal plane divides the body into right and left halves; the transverse plane divides the body into anterior and posterior halves; and the frontal plane divides the body into dorsal and ventral halves.

the history of this evolution while, for the most part, remaining a rather conservative skeleton when compared with those of other mammals. For example, primates possess a functional clavicle between the sternum and scapula; the tibia and fibula of the lower leg are separate bones (except in *Tarsius*, where they are fused for about the lower half of their length); the hands and feet have five digits (pentadactyl); there are no major fusions or reductions of limb bones; the seven cervical vertebrae have been retained; and there have been no major structural specializations in the skull, as for horns or tusks.

POSTCRANIAL SKELETON

The postcranial skeleton refers to the skeleton behind or, in the case of humans, below the skull (fig. 7.2). All bones of the postcranial skeleton develop endochondrally except the clavicle, which is a dermal bone as discussed in chapter four. Surface markings such as ridges, tubercles, and crests are common on many of these bones, indicating the former position of muscle attachments during the life of the animal. Knowledge of these surface markings and the muscles and tendons responsible for them (musculo-skeletal anatomy) may give useful information regarding the locomotor habits of their former owners.

The internal or microscopic structure of bone also reveals information regarding how the bone was used during the lifetime of an animal. For instance, a long bone, such as the human femur (fig. 7.3 a and b), has a compact cortical layer of bone of various thickness along its shaft, while the ends of the bone consist almost entirely of cancellous (spongy) bone with only a thin compact layer. The cancellous bone is arranged into irregular tracts called trabeculae, which project from the compact layer into the marrow cavity. These tracts coincide in all bones with lines of stress and are designated as trajectories. The lines of stress counteracting the forces applied to the human femur (compression and tension) curve from the upper and lower surfaces of the head and neck to enter the compact layer of the shaft. As shown in figure 7.3 a and b, there are two major systems of trajectories crossing each other at right angles which are then arranged around the neutral axis. The interdependence of structure and function, as represented by the human femur, is one of the basic principles of biology.

The primate forelimbs and hindlimbs have somewhat different functions which are reflected in their skeletal anatomy. Because primates have been primarily arboreal creatures since their origin, they have

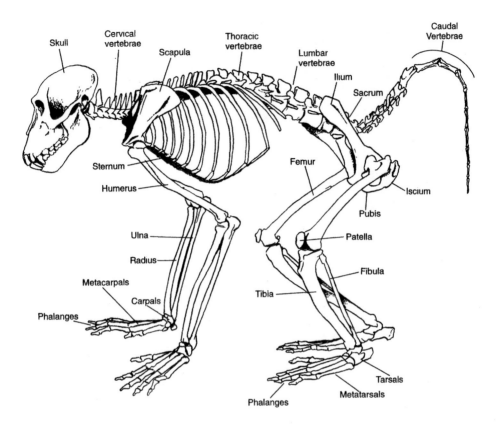

7.2 Primate skeleton (baboon) showing the major bones or groups of bones.

never developed the musculo-skeletal anatomy of a completely terrestrial animal. Consequently, the forelimbs are more adapted for grasping and exploration, although support and suspension are also part of their functional repertoire. The hindlimbs, on the other hand, support the body weight of the animal and are used in propulsion. Thus, the primate shoulder girdle (shoulder joint) is designed more for mobility, while the pelvic girdle (hip joint) is for support and stability.

The Forelimb

The shoulder girdle, also known as the pectoral (breast) girdle, consists of two bones, the clavicle and the scapula. The clavicle passes between the sternum and the scapula, connecting the skeleton of the forelimb with that of the trunk. Its chief function is that of a strut, holding the forelimb away from the side of the trunk and permitting a wide range of movement at the shoulder joint. The S-shaped clavicle is also known as the collar bone in humans, and is found in all primates (fig. 7.4).

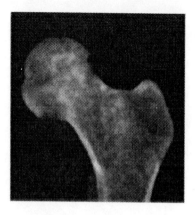
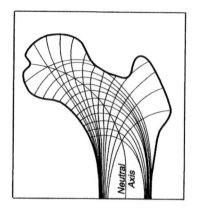

7.3 (left) Cross section of the proximal end of the human femur displaying the organization of cancellous and compact bone (Photo courtesy of Lee Haines) (right) A drawing of the same section showing the organization of trajectories around the neutral axis.

The scapula, or shoulder blade, is basically a triangular bone with three borders of varying length (fig. 7.5). A scapular spine passes across the dorsal surface of the scapula, and in the region of the shoulder joint it expands into the broad, flat, acromion process. The acromion is present in all primates, but it is larger in the anthropoid apes and humans. The scapula's shape and proportions are quite sensitive to different locomotor habits and consequently vary a great deal among primates (fig. 7.5). In the more quadrupedal primates, such as baboons, the scapula has a vertebral border (the border facing the vertebral column) that is short relative to the other borders (fig. 7.2), whereas in the more arboreal forms, including the anthropoid apes and humans, the vertebral border is relatively long in proportion to the other borders. The increase in the length of the vertebral border of the scapula allows for more effective leverage in rotating the scapula and elevating the arm to its fullest extent.

The shoulder joint is a ball-and-socket joint. The ball is the rounded, spherical head of the humerus (arm bone), and the socket is the shallow, concave surface on the lateral end of the scapula, the glenoid cavity. It should be noted that primates can move the humerus in its socket, which also moves, since it is part of the mobile scapula moving and rotating on the underlying rib cage. The result is an extremely flexible joint.

There is one bone in the upper arm, the humerus (fig. 7.6). It is a rather robust bone and must support a considerable amount of body

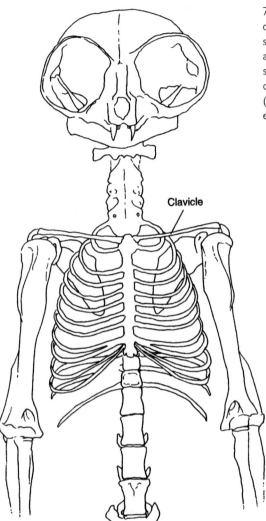

Clavicle

weight in quadrupedal primates. The rounded head of the humerus is well formed in all primates and displays various degrees of medial rotation among primates, from the more quadrupedal forms, where it points backward, to the more arboreal species, where it faces medially. This humeral torsion, or twisting, is a result of the different modes of locomotion that have evolved among living primates. The primate humerus may have a structure that is a holdover from a basic mammalian condition; that is, an entepicondylar foramen is present on the medial side of the lower end of the humerus which transmits an artery and nerve (fig. 7.6). This foramen is present in all but two prosimians, *Arctocebus*

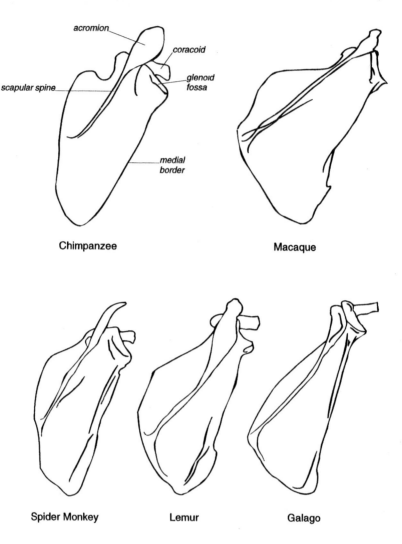

7.5 The lateral aspect of the right scapula in five primates. Specimens reduced to the same length.

and *Perodicticus*, although it may occasionally be present in these two lorises (Ankel-Simons 1983). It is variably present in platyrrhines but is very rare among catarrhines, especially anthropoid apes and humans.

The distal end of the humerus presents a globular capitulum and a concave trochlear which articulate with the radius and ulna respectively. It is here, at the elbow joint, that the various motions of flexion, extension (humerus and ulna), and rotation (humerus and radius) of the forearm on the arm occur (see below). The high degree of rotation of

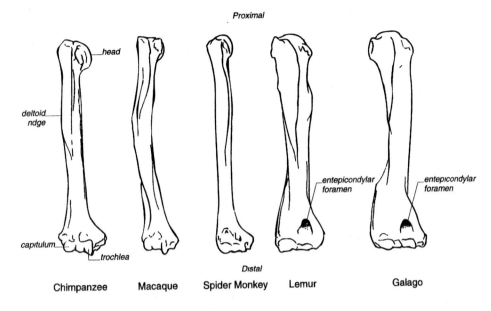

head

deltoid
ndge

entepicondylar
foramen

entepicondylar
foramen

capitulum

trochlea

Distal

Chimpanzee   Macaque   Spider Monkey   Lemur   Galago

7.6 The anterior aspect of the right humerus in five primates. Specimens reduced to the same length.

the forearm on the arm is an important function in all primates. Among other things, it permits the animal to pick up, manipulate, and examine objects.

The forearm has two parallel bones, the radius on the thumb (pollex) side, and the ulna on the little finger side (figs. 7.7 and 7.8; also see fig. 7.2). The two bones are always separate in primates. The radius is more robust than the ulna, and both bones are usually fairly straight, although some curvature of both bones may be found among primates. The head of the radius is almost circular and is well adapted for free rotary movements with the rounded capitulum of the humerus. The rotation of the radius around the ulna is called pronation when the thumb and radius rotate to the inside of the forearm (nearer the body) and supination when the thumb and radius lie to the outside of the forearm. The amount of supination is always less in quadrupedal forms than it is among hominoids. The olecranon is elongated in all primates, particularly in some of the more terrestrial species such as *Papio*. It is less projecting in the hominoids, permitting a more complete extension at the elbow joint.

The wrist (carpus) and hand (manus) of primates are made up of many separate bones (fig. 7.2). The carpal bones are arranged in two

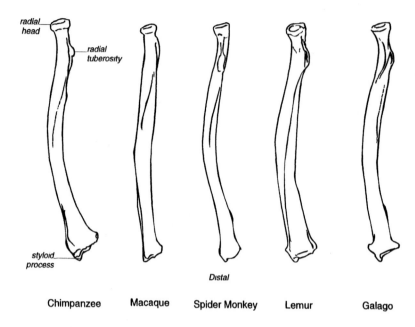

7.7 The anterior lateral aspect of the right radius in five primates. Specimens reduced to the same length.

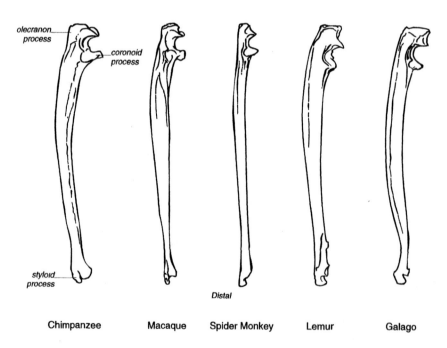

7.8 The anterior medial aspect of the right ulna in five primates. Specimens reduced to the same length.

rows, one proximal and one distal, and in most primates there are nine carpal bones forming the wrist. The os centrale is present in all primates; however in hominoids it is generally present only during early development. In gibbons and orangutans the os centrale remains a separate bone until well into adult life, when it may fuse with the scaphoid. Among gorillas and chimpanzees it fuses with the scaphoid in late fetal life. In humans the os centrale is present in early fetal life but merges with the scaphoid sometime during the third month of gestation.

The metacarpal bones (palm bones) are long, narrow bones with enlarged ends (fig. 7.2). In most primates the metacarpals are long compared to the phalanges (finger bones), although in some species of Old World monkeys and prosimians they are shorter than the proximal (first) phalanges. There are three phalanges per finger but only two for each thumb. The relative length of the fingers can be written as a digital formula, III > IV > II > V > I. This formula holds for all primates but lemurs, where the fourth digit is the longest. In addition, in *Arctocebus* and *Perodicticus* the index finger is very short and has only the remnant of the first phalanx left, while the third finger is shorter than the fourth finger (see fig. 2.5). The result is an extremely wide span for grasping branches or an excellent adaptation for the "slow, creep-up-and-grab" style of hunting employed by these lorisines (Kavanagh 1984, p. 60). The pollex is variably developed among all primates. True opposability of the thumb (the ability to pass the thumb across the palm while rotating it around its longitudinal axis) is found only in Old World monkeys, apes, and particularly in humans. In platyrrhines, the thumb is usually rather short, while in the spider monkey, *Ateles*, it is completely absent except for the stump of the metacarpal bone (fig. 7.9). Among Old World monkeys it varies in length from a small, atrophied nodule (*Colobus*) to a relatively long, well-formed structure in *Macaca* and *Papio*. As noted by Schultz (1972, p. 87), the lengthening of the fingers and the absence of the thumb have modified the hands of these "expert brachiators into mere grappling hooks." Interestingly, the thumb is shortest in the most arboreal and acrobatic of the great apes, the orangutan of Borneo and Sumatra.

The African great apes, though brachiators in the trees, proceed by knuckle-walking while on the ground, that is, they place the back side of their middle and terminal phalanges of fingers II–V on the ground while walking (see fig. 2.32b).

Primates generally have nails on their terminal phalanges rather than claws. Claws consist of two layers, a relatively thin superficial layer and

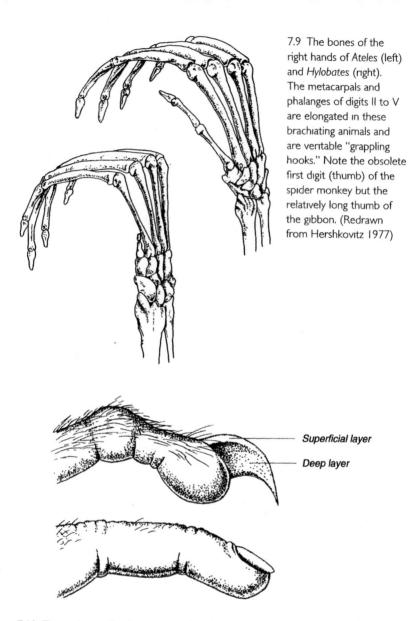

7.9 The bones of the right hands of *Ateles* (left) and *Hylobates* (right). The metacarpals and phalanges of digits II to V are elongated in these brachiating animals and are veritable "grappling hooks." Note the obsolete first digit (thumb) of the spider monkey but the relatively long thumb of the gibbon. (Redrawn from Hershkovitz 1977)

Superficial layer

Deep layer

7.10 The anatomy of a claw compared to that of a nail. The deep layer is absent in the nail.

a somewhat thicker deep stratum (fig. 7.10). The deep stratum is present in the claws, at least to some extent, in the aye-aye and marmosets (Le Gros Clark 1971). Nails, on the other hand, have retained only the superficial layer. The aye-aye has sharp claws on all of its fingers, while all other lemurs possess nails. Tarsiers also have small, triangular nails on all of their fingers. Among ceboids, marmosets have sharp, recurved claws on all digits of the manus while all other New World monkeys have nails. In Old World monkeys all fingers possess flattened nails as they do among hominoids.

The Hindlimb

The pelvic girdle (figs. 7.11 and 7.2) is structurally quite different from the pectoral girdle. As noted above, the pectoral girdle is adapted for mobility while the pelvic girdle is adapted for stability. There are three separate bones (ilium, ischium, and pubis) that join to form one side of the pelvis (os innominatum, which means, "a bone without a name") usually sometime prior to adulthood in primates. The ischium is enlarged into the ischial tuberosity, which forms a sitting platform that is particularly well developed in Old World monkeys since the ischial tuberosities support the ischial callosities or "sitting-pads" in these species (fig. 7.12). The ilium articulates with the sacrum (fused vertebrae of the spinal column), and the two pubic bones form a secure symphysial joint in front. In primates, the sacrum is composed of anywhere from two to nine fused sacral vertebrae; the average is three for the majority of species (Schultz 1961). The hip socket, or acetabulum, is deep and receives the head of the femur. This is also a ball-and-socket joint allowing movements in many directions.

The femur is the single bone of the thigh and has an articular head and two or three projections (trochanters) on its upper (proximal) end (fig. 7.13). The shaft is generally straight in most primates except in the hominoids, where it is bowed from front to back, especially in the great apes. If a third trochanter is present, it projects from the upper lateral border of the femur somewhat below the greater trochanter. Third trochanters are present on the femurs of prosimians, including tarsiers, present in callitrichids, but rare in other New World monkeys, and almost always absent among Old World monkeys, apes, and humans. The distal end of the femur has two condyles, a deep intercondylar groove or notch between them, and a groove for the patella (knee-cap). The condyles articulate with the tibia where flexion and extension occurs between the thigh and lower leg.

There are two bones in the lower leg, the tibia and fibula (not shown in fig. 7.14, but see fig. 7.2). These bones remain separate in all living primates with one exception, *Tarsius*. The tibia is larger and more powerful than the fibula and, as noted above, is the sole articulation with the femur. The fibula (shown only in fig. 7.2) is a slender, rather splint-

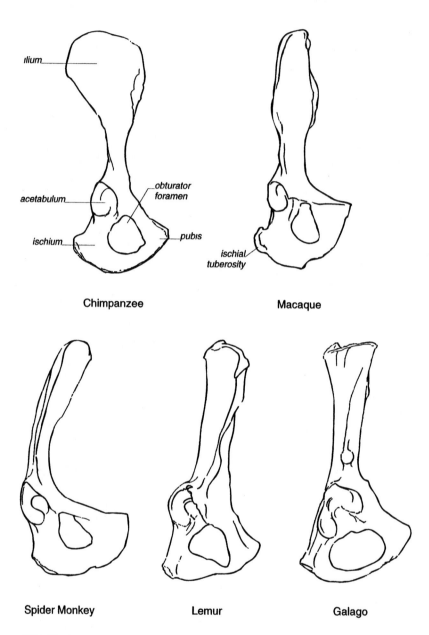

Chimpanzee

Macaque

Spider Monkey

Lemur

Galago

7.11 The lateral aspect of the right pelvic bone in five primates. Specimens drawn to the same height.

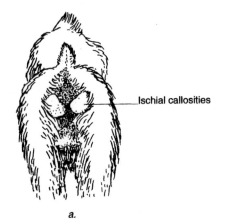

7.12 a) Ischial callosities of a male baboon. b) Baboon sleeping in a tree resting on its ischial callosities (sitting pads). (Redrawn from Napier and Napier 1985)

Ischial callosities

a.

b.

like bone presenting proximal and distal surfaces that articulate with the tibia. Incidentally, the tibia is located to the front and inner side of the lower leg, and its anterior border is what we occasionally bump when we trip as we run up the stairs. Both tibia and fibula end at the ankle.

In tarsiers the tibia is almost as long as the femur, and the fibula is fused to it for about the lower half of its length. The tibio-fibular fusion represents an adaptation to the locomotor habits of tarsiers, that is,

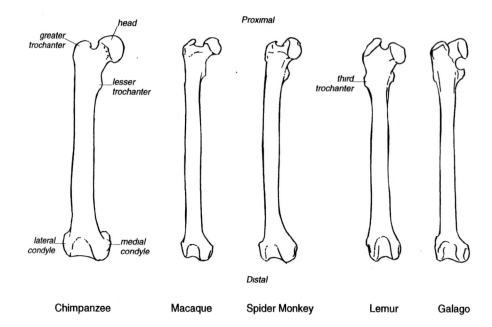

7.13 The anterior aspect of the right femur in five primates. Specimens reduced to the same length.

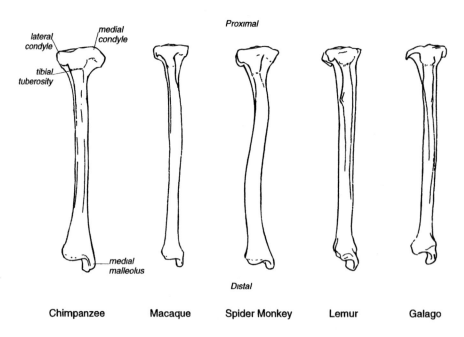

7.14 The anterior aspect of the right tibia in five primates. Specimens reduced to the same length.

they are superb leapers capable of leaping up to 6 meters (20 ft)—not bad for an animal that weighs only about 130 grams (4.5 oz).

Like the hand, the foot (pes) is composed of three different classes of bones, which may vary a great deal in size and proportions depending on the mode of locomotion (fig. 7.2, also see fig. 2.5). There are seven tarsal bones that have undergone various structural modifications during the evolution of the primates. The talus articulates with the tibia and fibula at the ankle, and the calcaneus forms the heel directly below the talus in all primates. The articulation at the ankle allows mainly extension and flexion. The most specialized foot is found in tarsiers, for here the tarsal bones are placed far forward by the elongation of the navicular and the front portion of the calcaneus (see fig. 2.5 C). These tarsal modifications are present to some degree in the saltatory (leaping) galagos, but the extreme elongation of the navicular and calcaneus relative to the other tarsal bones is found only in *Tarsius*. Indeed, the genus name *Tarsius* comes from the animal's highly specialized tarsal bones.

The big toe (hallux) is set apart from the other toes and forms a grasping organ in nonhuman primates (fig. 2.5 A). The grasping power of the primate hallux is greatly enhanced in the more arboreal species. A characteristic of the lemur foot is its large hallux with its wide separation from the second toe; the longest toe is always the fourth. The human foot is specialized away from the grasping foot found in all other primates. The big toe is not abducted from the other toes but is parallel with them; thus, the human foot is unique among primates since it is the only foot lacking a grasping big toe (see fig. 2.5 D). The human big toe is occasionally the longest toe in the foot. This is apparently the result of reduction of digits II-V rather than hypertrophy of the big toe, an observation made over forty years ago by Schultz (1956).

Lemurs have nails on their terminal phalanges, with the exception of the second, which bears a claw commonly known as the "toilet-claw" used for scratching and cleaning functions. The aye-aye has claws on all toes except the hallux, which possesses a nail. Among New World monkeys nails are present on all pedal digits except in callitrichids, where sharp, curved claws are present except on the hallux, which bears a nail. Nails are present on all pedal digits in Old World monkeys, apes, and humans.

Vertebral Column and Thorax

The vertebral column is composed of individual vertebrae divided into five morphological and functional regions: cervical (neck), thoracic (chest), lumbar (loin), sacral (sacred), and caudal (tail) regions (fig. 7.2). The total number of vertebrae varies widely among primates. Adolph Schultz (1961) studied the vertebral columns of 1,884 primate specimens from which the following information is taken. There are always seven cervical vertebrae in primates, which incidentally is the typical number in mammals. The greatest number of thoracic vertebrae, thirteen to seventeen, is found in lorisids. Several of the higher forms of New World monkeys have fourteen or fifteen thoracic segments, while among Old World monkeys twelve (and sometimes thirteen) is the usual number. In gibbons and African apes the numbers vary between twelve and fourteen. In orangutans it is between eleven and thirteen. Humans almost always have twelve thoracic vertebrae.

The number of lumbar vertebrae varies from three, found only in apes, to ten in *Lepilemur*. Platyrrhines usually have between four and five lumber segments while the catarrhine monkeys have seven. The shortest lumbar regions are present in hominoids who on average have five (in gibbons and humans), four (in orangutans) or only 3.6 (in African apes). It is easily understood why the African apes have such limited lateral movement within their broad trunks while the narrow trunks of Old World monkeys have so much lumbar flexibility.

The sacral region was discussed previously with the pelvic girdle. The caudal or tail vertebrae form the region most variable in number since tail length varies from being very long in some species (New World monkeys) to very reduced (Lorisidae) or even completely absent (gibbons, great apes, and humans). It is apparent that tail length has been reduced independently several times during the history of the primates.

The thorax or rib cage of primates varies with respect to the posture and locomotion of the animal. It is narrow, deeper than it is wide, and is suspended beneath the vertebral column in lower primates and the more quadrupedal forms. The distance between the thorax and the pelvis is great due to the increase in the number of lumbar vertebra. In hominoids, especially humans, the chest is much broader than deep and the space between the pelvis and thorax is much reduced, as are the number of lumbar vertebra. Commensurate with these structural changes in the thorax there has been a repositioning of the shoulder girdle. Thus, in monkeys the shoulder girdle is positioned on the side of the thorax and the shoulder joint lies opposite the sternum, whereas in hominoids the shoulder girdle has shifted onto the back of the chest wall and the shoulder joint is on a plane with the vertebral column. One consequence of these skeletal modifications has been an increase in the flexibility of the vertebral column, which is particularly evident in the long lumbar region of monkeys, in contrast to the more limited movement of the short lumbar area of hominoids. Indeed, in the African great apes the distance between the lower ribs and the upper border of the ilia is so short that they almost touch.

LOCOMOTION

Locomotion is defined as the act of moving from place to place, but locomotion also includes hanging and sitting postures, since these behaviors are also related to the anatomy of the trunk and limbs. Primates are essentially tree-living animals that have, in several instances, become terrestrial during their history. Of the various orders of mammals alive today, primates have evolved the greatest diversity of locomotor habits. The different locomotor habits have been variously defined and classified through the years, but, as with all classifications, it is difficult to assign specific locomotor labels to particular species since there is frequently overlap of locomotor behavior among them. In other words, most primates today have the capacity for using several types of locomotion, which makes it difficult to pigeonhole them into specific categories. During the past decade or so there have been numerous studies indicating the artificial nature of primate locomotor classifications. A similarity in morphological structure among several species does not always indicate a similarity in function. Nevertheless, classifications offer a framework for further research and discussions. Table 7.1 presents a locomotor classification of four major categories (with several subdivisions) originally proposed by Napier and Napier in 1967. This classification has been widely adopted through the years and represents only one of several approaches for beginning to understand the complexities of primate locomotion. Perhaps the weakest category in this classification is "semibrachiation." Mittermeier and Fleagle (1976), for example, demonstrated that *Ateles geoffroyi* and *Colobus guereza*, although considered semibrachiators, have very different types of locomotion. Furthermore, these authors correctly point out that it is dangerous to attribute the locomotor habits of a segment of a species to the whole species. Napier and Napier (1985, p. 45) also recognized the inaptness of "semibrachiation" but decided to retain the term "in the interest of stability."

One of the major differences among primates belonging to these respective locomotor categories is the relationship between the lengths of

Table 7.1   Classification of Primate Locomotor Types

Category	Sub-Type	Activity
Vertical clinging and leaping		Vertical clinging in and between trees. Hopping on the ground
Quadrupedalism	Slow climbing	Cautious climber No leaping
	Branch running and walking	Climbing, springing, and jumping
	Ground walking and running	Tree-climbing, rock climbing
	New World semibrachiation	Arm-swinging with use of prehensile tail
	Old World semibrachiation	Leaping and arm-swinging
Brachiation	True brachiation	Arm-swinging Bipedal branch running
	Modified brachiation (Orangutan)	Climbing and swinging using the four limbs
	(Chimpanzee)	Occasional arm-swinging, knuckle-walking on the ground
	(Gorilla)	No adult arm-swinging, knuckle-walking on the ground
Bipedalism		Standing, walking, and running

Adapted from Napier and Napier 1985

their forelimbs and their hindlimbs (fig. 7.15). These different proportions are expressed as the intermembral index:

$$\frac{\text{length of humerus} + \text{length of radius}}{\text{length of femur} + \text{length of tibia}} \times 100$$

The intermembral index ranges from about 53 to 150 in living primates. Although correlated with different locomotor habits, categoriza-

7.15 Three principal locomotor categories of primates showing the relative relationships between the forelimbs and hindlimbs. Upper: vertical clinging and leaping, middle: quadrupedal walking; lower: brachiation. (Modified from Napier 1977)

tion should always be based on observed behavior (Napier and Napier 1985).

Vertical clingers and leapers are all prosimians with long hindlimbs compared to their short forelimbs (the mean index ranges from 53 to 64). Bush babies and tarsiers have the longest legs relative to their arms, with a mean intermembral index of 53 and 56 respectively. Quadrupedal primates are mostly monkeys with forelimbs and hindlimbs of about equal length (mean index from 67 to 106). The quadrupedal category contains two prosimians, Lemur catta (67) and L. variegatus (72). The brachiators are all apes and have long forelimbs and relatively short hindlimbs (mean index from 108 to 145). Bipedalism is found only in humans, whose legs are somewhat longer than their arms (mean index around 70).

Primates live in forest or ground habitats, which are divisible into several vegetational settings including tropical rain forests, tropical mountain forests, mixed deciduous forests, woodland and open savannas, and arid scrub regions. The tropical forests are typically divided into four layers or stories: shrub layer, under story, middle story, and upper story (fig. 7.16). The majority of primates live in the under and middle sto-

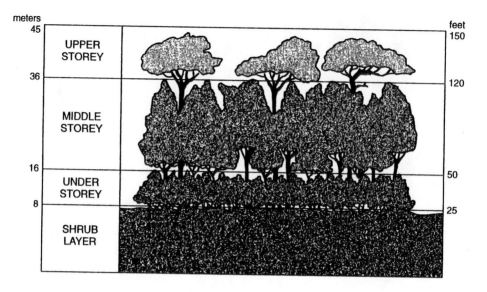

```
meters                                                              feet
45                                                                  150

        UPPER
        STOREY

36                                                                  120

        MIDDLE
        STOREY

16                                                                  50
        UNDER
        STOREY
8                                                                   25

        SHRUB
        LAYER
```

7.16 The different habitation levels of a tropical forest. A level may be occupied by a single primate species or by several species at the same time. (Napier and Napier 1985)

ries, where eating and traveling is rather easy compared with the upper story. Most traveling is quadrupedal, although some brachiation and leaping may be employed. Primates living in the upper layers, which may reach a height of 150 feet (45m), must leap or swing from tree to tree since many of the tree tops are separated from each other at this height. Many species inhabit the grasslands and woodland savannas, for example, baboon troops and patas monkeys. Mountain gorillas have adapted to life in the high montane forests at altitudes between 2,000 and 3,000 meters (6,000–10,000 ft).

Vertical Clingers and Leapers

This mode of progression is found in African bush babies (*Galago*), Malagasy lemuroids (*Avahi, Lepilemur, Propithecus, Hapalemur,* and *Indri*), and Southeast Asian tarsiers (*Tarsius*). It is characterized by rapid saltation (leaping) through the trees. The main thrust for movement is in the hindlimbs, which are quite long compared with the short forelimbs (fig. 7.15). The feet are prehensile and firmly grasp the trunk of a tree when landing or in the resting position. The hands are used for grasping and have little or no role in propulsion. This mode of progression can be quite rapid as the animal leaps from one tree to another as it makes its way through a forest. The tail, if present, may be used as a prop or for balancing, as in tarsiers. When on the ground the animals progress by bipedal hopping, or they may assume a quadrupedal stance when moving more slowly across the ground.

Quadrupedalism

This is the largest locomotor category in table 7.1. Napier and Napier (1985) recognize five subdivisions within this class. The fore- and hind-limbs in quadrupedal animals are of nearly equal length and, whether in the trees or on the ground, their gait involves the alternate use of each of the four limbs (fig. 7.15). Thus, the right forelimb and left hindlimb move together followed by the left forelimb and right hindlimb. Primates in the various subdivisions emphasize different anatomical regions as they locomote, but the basic mechanical plan remains essentially the same whether climbing, leaping, or arm-swinging.

Quadrupedal climbers lift the body by the forearms in a hand-over-hand fashion followed by the hindlimbs. This type of locomotion is practiced by the New World marmosets and capuchins, the Old World *Cercopithecus*, and some Malagasy lemurs.

Leaping is found in the Old World semibrachiating, leaf-eating monkeys of the genera *Colobus*, *Presbytis*, and *Nasalis*. It is interesting to note that when these monkeys land after a leap they employ their forelimbs to contact the branch first, or at least at the same time as their hindlimbs, which is just the opposite in the vertical clingers and leapers (Napier and Napier 1985). New World semibrachiators generally drop or plunge rather than leap from one perch to another. They also have prehensile tails for suspensory locomotion as well as for functioning as a fifth hand.

Terrestrial quadrupedalism is the mode of locomotion of several species of Old World monkeys, such as the savannah-dwelling baboon (*Papio*), the patas monkey (*Erythrocebus*), the gelada baboon (*Theropithecus*), and many macaques (*Macaca*). All of these primates, however, are good tree-climbers and run for the trees for safety when danger is near, or, as in the case of geladas, head for the rocky cliffs. They also sleep in trees or cliffs at night. These animals employ a digitigrade posture, that is, the body weight is carried through the distal half of hands and feet (the phalanges are parallel with the ground).

Several members of the Lorisidae (*Loris*, *Perodicticus*, *Nycticebus*, and *Arc-*

tocebus) progress by slow, deliberate hand-over-hand and foot-over-foot movements along the substrate. As mentioned before, hand and foot modifications are toward a forcepslike grasping mechanism. These animals never leap.

Brachiation

This specialized mode of locomotion is characterized by arm-swinging (figs. 7.15 and 7.17). The forelimbs are much longer than the hindlimbs and the hands have elongated digits II to V while the thumb is short relative to the other digits. True brachiation is practiced by gibbons and siamangs as they locomote through the trees of the Malayan rain forest. As a gibbon brachiates along a rod in figure 7.17, notice that the legs become nearly fully extended as it swings downward and forward and the body rotates forward. The legs are then brought up under the body and the body continues to rotate as the animal reaches with its free arm for the next grasp of the rod. The different positions of the trunk help the animal propel itself through the trees. Their speed has not been clocked, but according to Kavanagh (1984) a gibbon can whip through the forest canopy at extremely high speeds and may hurtle through the air between branches for as much as 15 meters (50 ft). He says that they "seem barely to tap the branches as they flash by" (p. 180). Gibbons also walk bipedally along branches or on the ground with their arms held upward and slightly backward for balance.

The great apes have a modified form of brachiation when in the trees. The orangutan is generally considered the best brachiator of the three, spending about 85 percent of its time in the trees. The chimpanzee is a better brachiator than the gorilla. The forelimbs of the chimpanzee are still much longer than the hindlimbs and play the more important role in propelling the body forward. The hands have short thumbs relative to the other digits, and the metatarsal and phalanges of digits II to V are curved. It is estimated that chimpanzees spend only about half of their time in trees during the day while gorillas are almost completely ground-living (Napier and Napier 1985). Both species sleep in trees at night.

When on the ground, orangutans walk on all fours and use clenched fists or the backs of their hands for support. The African apes, as mentioned earlier, are knuckle-walkers. The hand is bent so that the body weight is carried through the knuckles, and since the arms are long the trunk is carried in a semi-erect position.

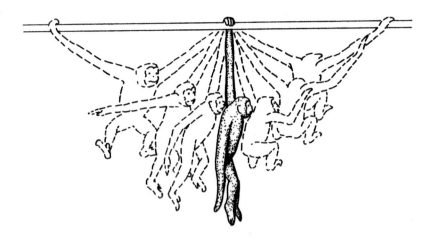

7.17 A gibbon swinging arm-over-arm along a rod demonstrating the use of the arms, legs, and body during brachiation. (Redrawn from Napier and Napier 1985)

Bipedalism

The majority of monkeys can assume a bipedal posture for only a short time, as when peering over tall grass or carrying objects in their hands. The chimpanzee and gorilla are capable of much longer periods of bipedality, although when on the ground they are normally quadrupedal and in the knuckle-walking attitude. True bipedalism is present only in humans. As Napier and Napier (1985, p. 50) stated, "The distinctive element in the human gait . . . is the ability to stride." No other primate can do this heel-toe walking in which the heel makes first contact with the ground. In other words, humans are the only bipedal primates.

It is obvious that living primates are difficult to categorize into specific locomotor habits since there is so much overlap in morphology, proportions, and functions. We are still learning much about the relationship of locomotion to body structure and body size, environment, diet, and social organization in living primates; and it is only through knowledge of the anatomy and function of living primates that we can begin to understand the locomotor habits of fossil primates.

8

Growth and Development

P. Mormon. Mandrill.

Primates, like all sexual animals, begin life as a fertilized cell that is the result of the union of two gametes (germ cells) and is known as a zygote. From the moment of conception the organism undergoes differentiation (i.e., growth and development), which continues throughout its life. The term "growth" usually refers to a change in the size of an organism or any of its parts from its earliest stage to maturity. "Development" is used in the sense of an increase in skill or complexity of function by which the embryo becomes a mature organism. Notice that these processes involve a series of changes, not just the addition of new material in order to become larger. Differentiation is the key to understanding how an organism grows and develops. The initial part of this continuum is concerned with the development of the embryo, which lasts but a few days or weeks. The embryo develops into the fetus, which continues to grow and develop until birth. The time this takes is quite variable among primates and is known as the prenatal or gestation period. Other things being equal, the length of gestation increases proportionally with increases in body weight and relative brain weight. The postnatal period commences at birth and lasts until the animal reaches adulthood, although it should be noted that biological changes continue throughout the life of an animal.

Certainly all members of the order Primates have an evolutionary connection relating them to each other, but since the major grades of the order have had a separate history for millions of years, "There is no evolutionary reason to expect that the patterns of growth of these divergent and ecologically distinct species should be identical or even similar" (Bogin 1988, p. 68). This chapter will consider some of the similarities and differences in primate growth patterns.

The changes (structural, physiological, and behavioral) that take place throughout ontogeny (development) occur at different rates and times in the various primate species and constitute an important area of primate research. Thus, growth may be: 1) a change in rate of growth per given unit of time; 2) a constant rate of growth but a change in duration; 3) a change in both rate and duration. Because evolutionary changes acquired by adults are the result of alterations in one or more of these processes, it is clear that ontogeny holds the potential for new variations that may lead to new selective advantages. Development is more than the process of converting an egg into an adult; it must ensure survival to reproductive age. As my colleague Dr. Linda Duchin has said, "In the past, these variations and advantages led to the development of our many primate relatives. Our kinship with these groups, and our link to a common ancestor, is seen perhaps most clearly in a study of embryology" (personal communication). In order to understand how primates grow and develop it is necessary to consider the structure and function of the organs responsible for these developmental processes.

External Genitalia

The external genital opening (vulva) of the female lies ventral or forward of the anus; the space between the two is the perineum. The genital opening expands into the vestibule, which receives the separate vaginal and urinary tracts. Just forward of the urinary opening is the clitoris, the homologue of the male penis. The clitoris is variable in size in female primates and is larger and more pendulous than the male penis in some prosimians (lemurs) and several New World genera (e.g.,

8.1 Sexual swelling around the tail and anal region of a female baboon. (Photo courtesy of Judy Johnson, Regional Primate Research Center, Seattle)

Cebus, Saimiri, Lagothrix, and *Ateles*), a feature that makes sex determination difficult in these genera. Among lorisform primates the urethra (urinary tube) passes through the clitoris opening at its tip, simulating the male condition, whereas in all other primates it opens at the base of the clitoris. In *Galago* and *Hapalemur,* the clitoris contains a baculum (a small rodlike bone). The labia majora and minora (external and internal tissue folds bounding the vulva) are also variable as to their presence and size among living primates. In general they undergo some atrophy during early development in most monkeys and great apes while remaining prominent in gibbons. They are present in humans.

A condition found only among the females of Old World primates is the sexual swelling of the anal and genital regions (sexual skin) during ovulation which, incidentally, coincides with the female's most receptive sexual period (fig. 8.1). In species of *Macaca, Papio,* and *Pan* the swellings may enlarge to include the areas around the base of the tail, lower back, and thighs. The sexual skin may turn red, particularly in chimpanzees. In some macaques and baboons the red area includes not only the turgescent area but also the skin over the thighs, chest, and face.

In all adult male primates the testes are permanently descended and occupy the scrotum. Their descent into the scrotum occurs at different ages but is generally earlier in the higher primates; in humans the testes normally descend late in fetal life. The testes are relatively large in most primates, being quite large in chimpanzees, whose testicles can weigh up to 250 grams (9 oz), whereas in gorillas they weigh only about 36 grams (Schultz 1972). The gorilla also has the smallest penis of the three great apes. It is known that the testes of some prosimians show seasonal changes, and Sade (1964) demonstrated that the testes of rhesus monkeys vary in size and are larger during their breeding season, which probably indicates an increase in sperm production. The scrotum lies behind the penis in all higher primates except the gorilla and gibbon, where it is on the sides of the penis. This is also its position in most prosimians and callitrichids. The penis is pendulous in all male primates, and except for tarsiers, spider monkeys, woolly monkeys, and humans, there is a baculum (os penis) in the distal part of the organ. There are marked differences in the shape and size of the baculum among the various species of primates, and Fooden (1971) suggests that they are useful as taxonomic indicators among macaques.

FEMALE REPRODUCTIVE ORGANS

The ovaries lie within the pelvic cavity and vary in size among female primates of various species. The complete supply of ova is formed before birth and gradually develops into mature eggs by the time the female reaches sexual maturity. Ovulation is spontaneous in primates, that is, the eggs are released from the ovaries without first having copulation. Thus, ovulation happens whether copulation occurs or not. Estrus (heat) usually occurs during ovulation in most female primates and is generally the time when the female is most receptive to the male. If the egg is not fertilized the wall of the uterus is shed along with bleeding (menstruation) in some primates. Strepsirhines do not show external bleeding, while platyrrhines, orangutans, and gorillas show little if any bleeding (Richard 1985). Old World monkeys, gibbons, chimpanzees, and humans all have blood with their menstrual discharge.

The genital tract is divided into the Fallopian (uterine) tubes, uterus,

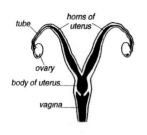
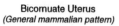

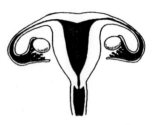

Bicornuate Uterus (General mammalian pattern)	Intermediate Type (Prosimians)	Unicornuate Uterus (Anthropoids)

8.2 The structure of a typical mammalian uterus (left) compared with the uterus of primates. (Redrawn from Napier and Napier 1985)

and vagina. The Fallopian tubes may be coiled or straight. Their free, open end receives the egg during ovulation. Normally conception occurs in the Fallopian tubes, after which the zygote (fertilized egg) passes through the Fallopian tube to become implanted in the wall of the uterus.

The uterus has a single body in all anthropoid primates, a condition that represents an early embryological fusion of two separate tubes and is known as a uterus simplex (fig. 8.2). In prosimians and tarsiers the tubes unite to form a single uterine body, but the upper parts of the tubes remain separate, resulting in what are known as uterine horns. This is also referred to as a bicornuate uterus.

The vagina is a muscular, highly dilatable canal lined with mucous membrane. It extends from the uterus to the external genitalia, where it opens to the exterior as the vulva.

REPRODUCTION

As mentioned above, fertilization normally occurs in the uterine tube, and the resulting zygote develops and passes into the uterus where it implants itself into the thickened lining of the uterus and becomes the embryo. The embryo is surrounded by and develops within two membranes, an inner amnion and an outer chorion. The amnion provides the embryo with an aqueous environment that protects it from outside pressures. The chorion forms the connection between the embryo and the mother's uterus and is the dominant fetal component of the placenta. The placenta is the most important structure during intrauterine development, carrying out such physiological activities as gaseous exchange, regulation of nutrients as well as other biological functions necessary for growth and development, and removal of waste products from the fetus. It also produces a variety of important endocrines. The structure of placental membranes has also proven very important in mammalian systematics because they are quite conservative. However, it remains true that many aspects of the organization of mammalian placentas probably have some functional significance. The physiological and systematic importance of these functional differences is thoroughly discussed by Martin (1990).

Placentas are classified in several ways. One method uses the chorionic outgrowths over the uterus surface. Primates have one of two types. The first type, diffuse, in which the chorionic protrusions are distributed over the surface of the placenta, is characteristic of all prosimians except *Tarsius*. The second type is known as a discoid placenta in which the chorionic villi are limited to circumscribed oval or round regions on the surface of the placenta. The discoid placenta is typical of all higher primates including tarsiers.

Another way of classifying placentas is the arrangement of the opposing fetal and maternal membranes. These opposing membranes form the placental barrier, or what King (1986) prefers to call the interhemal (blood) membrane. The basis of this classification depends on the reduction in the number of maternal membranes, since the number of fetal membranes remains the same (fig. 8.3). There are several arrangements described for mammals, but only two types are present in primates. The epitheliochorial placenta has contact between the uterine epithelium and the fetal chorion with no reduction in maternal membranes. This type of placenta is common to all prosimians except tarsiers. The hemochorial placenta has lost the maternal membranes and has only the chorion. In addition, the maternal blood vessels have broken down so that the chorion is directly bathed in maternal blood. All anthropoid primates and tarsiers have this type of placenta. The epitheliochorial placenta is more primitive and less efficient, that is, it is a greater barrier to the transport of nutrients and oxygen than is the hemochorial placenta, which provides enhanced food and oxygen supply to the fetus as well as more rapid elimination of waste products.

After birth the placenta may separate from the uterus without loss of the uterine wall or maternal blood. This is known as a nondeciduate placenta. If, however, maternal tissue and blood is lost with the placenta, it is referred to as a deciduate placenta.

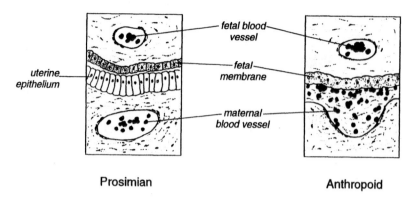

8.3 Diagram showing the relationship of the fetal and maternal membranes and blood vessels in prosimians and anthropoids. (Modified from Hamilton et al. 1957)

The subject of primate placentation is a vast and complicated one and has only been touched on here. The major differences among primate placentas are as follows: lemurs and all lorisids have a diffuse, nondeciduate, epitheliochorial placenta. An exception is Demidoff's galago (*Galago demidovi*), which has a more complicated placental morphology (see Ankel-Simons 1983). The anthropoid primates as well as tarsiers have a discoidal, deciduate, hemochorial placenta.

PRENATAL PERIOD

The number of offspring that will be born to primate mothers is usually one. Marmosets, however, normally have twins, and multiple births are not uncommon in prosimians. In higher primates, including humans, twins happen in about one out of a hundred births, and approximately 80 percent are fraternal (derived from two eggs) (Ankel-Simons 1983). Prenatal development is divided into three main periods: 1) the period culminating with the implantation of the zygote but before the establishment of intra-embryonic circulation, during which the fetal membranes are established and the germ layers are formed; 2) the embryonic period, a time of rapid growth and differentiation, during which all of the main systems and organs of the body and the major features of the external body form are established; and 3) the fetal period, beginning at the end of the embryonic period and lasting until birth. These periods differ in length among primates, and certainly more is known about early human development than about that of many other primates. In humans, for example, the embryonic period

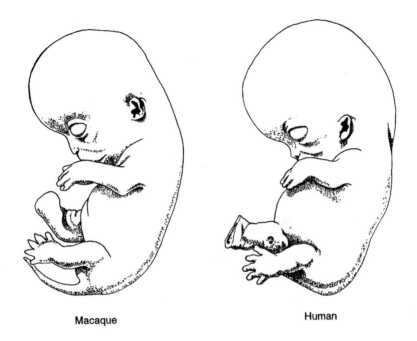

Macaque   Human

8.4 Fetuses of a macaque and human of approximately the same age (44 and 49 days respectively). (Redrawn from Schultz 1972)

lasts from the beginning of the fourth week to the end of the eighth week. The fetuses of a macaque and a human are shown at about the same embryonic age in figure 8.4. They are approximately the same size, and except for the tail in the macaque, they are almost identical in external appearance.

The duration of the prenatal period varies a great deal among primates (table 8.3). For example, in Microcebus the prenatal growth period lasts about two months, in lemurs it varies between two and five months, in most platyrrhines between four and one-half and four and three-quarter months, in Old World monkeys between five and six months, and in the great apes between eight and nine months. The great apes, therefore, have a gestation period comparable to that of humans. Gibbons, although smaller than baboons, have a longer gestation period. New World monkeys, though there are differences in their body sizes, all have approximately the same period of gestation. Additionally, the size of the birth canal and the size of the terminal fetus is a critical relationship among primates as we shall shortly see. One thing seems clear among primates: there is no close correlation between adult body size and duration of the gestation period.

Table 8.1 Vital Statistics of Living Primates

Taxa	Female Body Wt. (Kg)	Male Body Wt. (Kg)	Neonatal Wt. (Grams)	Approximate Gestation (Days)	Neonatal Brain Weight (Grams)
Lemuridae					
Lemur catta	2.50	2.90	88.2	135	8.8
L. fulvus	1.90	2.50	81.4	118	10.7
L. macaco	2.50	2.50	100.0	128	—
L. mongoz	1.80	1.80	—	128	—
Varecia variegatus	3.10	3.60	107.5	102	10.6
Hapalemur griseus	2.00	2.00	48.0	140	—
Lepilemuridae					
Lepilemur mustelinus	0.64	0.61	34.5	135	2.9
Indriidae					
Indri indri	10.50	10.50	300.0	160	—
Avahi laniger	1.30	1.30	—	—	—
Propithecus verreauxi	3.50	3.70	107.0	140	—
Cheirogaleidae					
Cheirogaleus major	0.40	0.40	18.0	70	—
Microcebus murinus	0.08	0.08	6.5	62	—
Daubentoniidae					
Daubentonia madagascariensis	2.80	2.80	—	—	—
Lorisidae					
Loris tardigradus	0.26	0.29	12.6	163	2.7
Nycticebus coucang	1.20	1.30	49.3	193	4.0
Arctocebus calabarensis	0.31	0.32	25.2	134	2.3
Perodicticus potto	1.08	1.02	46.5	193	—
Otolemur crassicaudatus	1.26	1.42	47.4	135	4.0
Galago demidovi	0.62	0.63	7.5	111	1.2
Galago senegalensis	0.21	0.24	11.5	124	2.3
Tarsiidae					
Tarsius spectum	0.20	0.20	30.0	157	—
Callitrichidae					
Callithrix jacchus	0.29	0.31	28.0	148	4.4
Saguinus midas	0.53	0.60	36.0	127	—
S. oedipus	0.51	0.45	43.2	145	4.9
Leontopithecus rosalia	0.55	0.56	53.6	129	—

Table 8.1   Vital Statistics of Living Primates (continued)

Adult Brain Wt. (Grams)	Life Span (Years)	Common Name	Diet[1]	Habitat[2]	Status[3]
25.6	27.1	Ring-tailed lemur	L, F	A, T	E
25.2	30.8	Brown lemur	L	A	E
25.6	27.1	Black lemur	L, F	A	E
21.8	—	Mongoose lemur	L, F, N	A	E
34.2	—	Ruffled lemur	F	A	E
14.7	12.1	Gentle lemur	Bs, L	A	E
9.5	—	Sportive lemur	L, Fl, B	A	E
34.5	—	Indris	L, F	A	E
10.0	—	Woolly lemur	L	A	E
27.5	—	Verreaux's sifaka	L, F	A, T	E
5.9	8.8	Greater dwarf lemur	F, I, G	A	E
1.8	15.5	Gray mouse lemur	F, L, I, Gum	A	E
45.2	—	Aye-aye	Gbs, Lar.	A, T	E
6.7	13.0	Slender loris	I	A	
10.0	14.5	Slow loris	F	A	
7.7	9.5	Golden potto	I, Gum	A	
14.3	10.0	Potto	F, G, Gum	A	
11.8	15.0	Thick-tailed bush baby	F, G, Gum	A	
2.7	14.0	Dwarf bush baby	I, F, G, Gum	A	
4.8	16.0	Lesser bush baby	I, G, Gum	A	
3.8	12.0	Spectral tarsier	Faun.	A, T	E
7.9	12.0	Common marmoset	F, I, G	A	
10.4	13.0	Red-handed tamarin	F, G	A	
9.0	13.0	Cottontop tamarin	F, Faun., G	A	E
12.9	—	Lion tamarin	F, Faun.	A	E

[1] L=Leaves   F=Fruits   N=Nuts   Bs=Bamboo Shoots   B=Buds   Gum=Gum, Sap or Resins
I=Insects   S=Seeds   G=Grass   Lar.=Larvae   Faun.=Fauna   Fl=Flowers   R=Roots   Gbs=Grubs
[2] A=Arboreal   T=terrestrial
[3] E=Endangered

Table 8.1   Vital Statistics of Living Primates (continued)

Taxa	Female Body Wt. (Kg)	Male Body Wt. (Kg)	Neonatal Wt. (Grams)	Approximate Gestation (Days)	Neonatal Brain Weight (Grams)
Cebuella pygmaea	0.14	0.16	16.0	136	—
Callimico goeldii	0.53	0.65	48.6	154	5.8
Cebidae					
Cebus apella	2.10	2.86	248.0	160	—
Saimiri sciureus	0.58	0.75	195.0	170	—
Alouatta palliata	5.70	7.40	480.0	187	30.8
Ateles geoffroyi	5.80	6.20	426.0	229	64.0
Lagothrix lagothricha	5.80	6.80	450.0	225	—
Aotus trivirgatus	1.00	0.92	98.0	133	10.1
Pithecia pithecia	1.40	1.60	—	163	—
Cercopithecidae					
Macaca fuscata	9.10	11.70	503.0	173	—
M. mulatta	3.0	6.20	481.0	167	54.5
M. nemestrina	7.80	10.40	473.0	167	66.0
M. sylvanus	10.0	11.20	—	—	—
Cercocebus albigena	6.40	9.00	425.0	177	—
C. galeritus	5.50	10.20	—	171	—
Papio cynocephalus	15.0	20.0	854.0	175	73.5
P. ursinus	16.80	20.40	—	187	—
Theropithecus gelada	13.6	20.50	464.0	170	—
Cercopithecus	3.56	4.75	314.0	163	—
C. mitis	4.40	7.60	402.0	140	—
C. neglectus	3.96	7.00	260.0	182	—
C. talapoin	1.10	1.40	180.0	162	—
Erythrocebus patas	5.60	10.0	—	163	—
Presbytis entellus	11.40	18.40	—	168	—
P. obscura	6.50	8.30	485.0	150	—
P. cristatus	8.10	8.60	—	—	—
Nasalis larvatus	9.90	20.30	450.0	166	—
Colobus polykomos	8.40	10.40	597.0	170	38.5
C. satanas	9.50	12.00	—	195	—
C. verus	3.60	3.80	—	—	—
Hylobatidae					
Hylobates lar	5.30	5.70	410.5	205	50.1
H. syndactylus	10.60	10.90	517.0	231	—
Pongidae					
Pongo pygmaeus	37.00	69.00	1,728.0	260	170.3
Pan troglodytes	31.10	41.60	1,756.0	228	128.0
Gorilla gorilla	93.0	160.0	2,110.0	256	227.0

Table 8.1   Vital Statistics of Living Primates (continued)

Adult Brain Wt. (Grams)	Life Span (Years)	Common Name	Diet[1]	Habitat[2]	Status[3]
4.2	10.0	Pygmy marmoset	G, I, Gum	A	
10.8	9.0	Goeldi's monkey	Faun., F	A	E
71.0	40.0	Tufted capuchin	I, F	A, (T)	
24.4	21.0	Squirrel monkey	I, F	A	
55.1	13.0	Mantled howler	L, F	A	
110.9	20.0	Black-handed spider monkey	F, L	A	
96.4	12.0	Common woolly monkey	F, L	A	
18.2	12.6	Night monkey	F, L	A	
58.2	15.0	White-faced saki	F, S	A	
109.1	22.0	Japanese macaque	F, L, Fl., I	T, A	
95.1	21.6	Rhesus monkey	F, L, Fl., I	T, A	
106.0	26.3	Pig tailed macaque	F, L, Fl., I	T, A	
93.2	—	Barbary macaque	F, L, Fl., I	T, A	E
99.1	21.0	Gray-cheeked mangabey	F, Faun. A		
114.7	19.0	Tana River mangabey	F, Faun.	T, A	
169.1	33	Yellow baboon	F, L, G, S, R	T, A	
214.4	—	Chacma baboon	F, L, G, S, R	T, A	
131.9	—	Gelada baboon	G, S, R	T	
59.8	31.0	Vervet monkey	f, Faun., G	A, T	
75.0	—	Blue monkey	F, I, L	A, T	
70.8	20	DeBrazza's monkey	F, I, L	A	
37.7	22.3	Talapoin monkey	I, F	A, T	
106.6	20.2	Patas monkey	S, G, F	T	
64.0	20.0	Hanuman langur	L, F, Fl.	A, T	
67.7	—	Dusky leaf monkey	L, F	A	
64.0	—	Silvered leaf monkey	L, F	A	
121.7	—	Proboscis monkey	L, F	A, T	E
76.7	26.0	Black-and-white colobus	L, F	A	
80.2	—	Black colobus	S, L, F	A	E
57.8	—	Olive colobus	L	A, T	
107.7	31.5	White-handed gibbon	F, L, Faun.	A	E
121.7	—	Siamang	L, F, Faun.	A	E
413.3	50.0	Orangutan	F, L, B	A, T	E
410.3	44.5	Common chimpanzee	F, L, Faun.	A, T	E
505.9	39.3	Gorilla	L, F	T, A	E

Data from Harvey, et al. 1987, Brizzee and Dunlap 1986, and Kavanagh 1983

Because there are differences in gestation periods as well as differences in the maturity of animals at birth, Martin (1975 and 1990) has described two major reproductive strategies employed by mammals: precocial and altrical. In the former condition the gestation period is relatively long, neonates are born in small litters, they are advanced in development with relatively large brains, their eyes are open, and infants usually have high mobility at birth. The infants develop more slowly with relatively long suckling periods, and the mother-infant bond is strong. There is a longer birth interval, resulting in a lower reproductive turnover. The populations tend to remain rather stable through time. This precocial pattern is known as the K-selected strategy. In mammals with the altricial complex, the other major reproductive strategy, gestation is relatively short, the young are born in large litters, they are relatively undeveloped, their eyes are closed, and they usually stay in their nests until self-sufficient. The infants mature quickly with a short suckling period, and the mother-infant bond is fairly weak. The interbirth intervals are relatively short and the reproductive turnover is high. Population size can fluctuate quickly with environmental changes. This pattern is known as the r-selected strategy.

In general, the K-selection strategy is favored in stable environmental conditions while the r-strategy operates in conditions of environmental fluctuations. The r-strategy is found, for example, among rodents, insectivores, and carnivores; the K-strategy is found in ungulates, cetaceans, and, typically, primates. Martin (1975) points out, however, that among primates there are several prosimian groups (Cheirogaleinae, Galaginae, aye-aye, and the variegated lemur) that display r-strategy altrical traits. Indeed, humans have been described as "secondarily altrical" since their central nervous system is still "embryonic" during the first year of postnatal life (Martin 1990).

Birth weights of several primate species are presented in table 8.1. It is obvious that there would be an increase in birth weight from prosimians to humans, but if we consider birth weight as a percentage of nonpregnant maternal weight we get the interesting results in table 8.2. It is immediately apparent that the smallest relative birth weights are found in the great apes, while the larger ratios are in the Old and New World monkeys. The ratio of the three prosimians is about that of the Old World monkeys. It seems that birth weight cannot exceed much more than 12 percent of the nonpregnant maternal weight. Birth size then becomes critical since it is so closely correlated with the size of the maternal birth canal, especially since the head is comparatively large in

Table 8.2   Birth Weight Expressed As a
Percentage of Nonpregnant Maternal Weight.

Species	Percentage
Microcebus murinus	6.0
Lemur fulvus	4.6
Galago senegalensis	5.0
Callithrix jacchus	12.0
Saimiri sciureus	14.7
Cebus capucinus	8.5
Alouatta palliata	8.0
Ateles geoffroyi	7.0
Macaca irus	10.0
Macaca mulatta	6.7
Papio hamadryas	7.1
Theropithecus gelada	10.0
Presbytis cristatus	7.0
Prebystis entellus	6.7
Nasalis larvatus	4.6
Hylobates lar	7.5
Hylobates syndactylus	6.0
Orangutan	4.0
Chimpanzee	4.0
Gorilla	2.6
Human	5.5

Adapted from Schultz 1972.

newborn primates. In fact, the relationship between head size and the space within the birth canal is a critical one in many primates, as shown in figure 8.5. It is obvious that the great apes have rather spacious birth canals in relation to the size of their heads at parturation compared to monkeys and humans. In fact, it has been suggested that the great apes could afford to continue their intrauterine period for a much longer time (Schultz 1972). In particular, the birth canals of macaques and humans are small compared to head size, which often results in protracted and difficult births, especially in macaques, who have achieved from 65 to 70 percent of their brain size at birth. In humans, whose brain at birth is from 25 to 30 percent of its adult size, the head is often turned to the side for easier passage through the birth canal. For a thorough

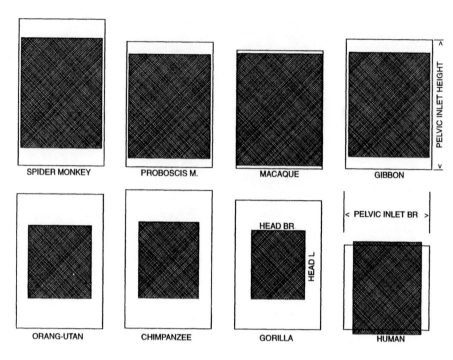

SPIDER MONKEY

PROBOSCIS M.

MACAQUE

GIBBON

PELVIC INLET HEIGHT

ORANG-UTAN

CHIMPANZEE

HEAD BR

GORILLA

HEAD L

< PELVIC INLET BR >

HUMAN

8.5 Diagram of the relationship between the dimensions of the pelvic inlet in adult female primates and the size of the head in newborns. (Redrawn from Schultz 1972)

review and analysis of the obstetrical problems of primates during birth, see Rosenberg and Trevathan (1996).

In addition to the obvious obstetrical functions of the primate pelvis, another factor that must be considered is the locomotor habits of the species (Leutenegger 1974). During the evolution of primate pelvic structure, locomotory functions as well as obstetrical demands have played important selective roles on the structure of the pelvis of living primates. Debates continue as to the relative importance of these and other influences, but it is clear that the evolution of the primate pelvis has been a complicated one and that it is far from being completely understood. As with many structural changes, the evolution of the primate pelvis probably represents a compromise of several different selective forces.

During prenatal and postnatal growth and development the animal undergoes continuous changes in shape and size which are mediated by alterations in timing, that is, variations in the rates and duration of growth during ontogeny. Such alterations in the timing and duration of events during ontogeny are called heterochrony (shifts in the

Table 8.3 Gestation Length in Primates

Species	Approximate Gestation Length (in months)
Loris tardigradus	5
Nycticebus coucang	3
Perodicticus potto	5 1/2
Galago senegalensis	4
Galago crassicaudatus	4 1/2
Galago demidovii	3
Microcebus murinus	2
Lemur catta	5
Propithecus verreauxi	5
Indri indri	2
Tarsius spectrum	6
Callithrix jacchus	4 3/4
Cebuella pygmaea	4 3/4
Saguinus oedipus	4 3/4
Callimico goeldii	4 1/2
Alouatta seniculus	4 3/4
Ateles paniscus	4 3/4
Lagothrix lagothricha	4 3/4
Macaca mulatta	5 3/4
Macaca fuscata	5 1/2
Macaca nemestrina	5 3/4
Papio cynocephalus	6 1/2
Theropithecus gelada	6
Cercocebus albigena	6
Cercopithecus ascanius	6
Cercopithecus aethiops	6 1/2
Presbytis entellus	5 3/4
Nasalis larvatus	5 1/2
Colobus polykomos	6
Hylobates lar	7 1/2
Hylobates syndactylus	7 3/4
Gorilla gorilla	8–9
Pongo pygmaeus	8–9
Pan troglodytes	8–9

Data from Brizzee and Dunlap 1986.

timing of developmental patterns). Shea (1981), for example, in a study of postnatal growth of three hominoids, has shown that the greater relative arm length of gorillas and chimpanzees compared to that of humans is due to the faster arm growth in the former animals compared with the slower arm growth of humans. Such heterochronic changes during ontogeny, whether prenatal or postnatal, have played important roles in the history of primates. The evolutionary biologist Steven Jay Gould (1977, p. 9) wrote, "If the frequency of heterochronic change were known, it would provide a good estimate for the importance of regulation as an evolutionary agent." The investigation of the integration of heterochrony and allometry (size and shape relations during ontogeny) during the period of growth and development has produced new insights and understanding of some of the most intriguing problems of morphological change confronting evolutionary biologists today (Gould 1977, Alberch et al. 1979, and Shea 1983). Details of these interesting and important studies are beyond the scope of this book, but excellent surveys can be found in McKinney (1988), Shea (1989), and Hall (1992).

The growth and development of fetal primates has been investigated in several groups of Old World monkeys and anthropoid apes, beginning with the pioneering work of Adolph Schultz in the 1920s (Schultz 1926). Schultz continued his studies of primates, particularly of their growth and development, almost until his death in 1976. He summarized his many contributions to primate biology with the publication of The Life of Primates in 1972. An important aspect inherent in studies of primate growth and development is the identification and analysis of ontogenetic changes (heterochrony/allometry) among living primates. Schultz was aware of the importance of ontogenetic changes when he wrote, "We must first collect the data on growth of other primates in order to demonstrate which ontogenetic changes represent peculiarities of man and which are common property of many, or even all, of the different primates" (Schultz 1937, p. 71).

The following discussion is based partly on the extensive review of nonhuman primate growth and development by Brizzee and Dunlap (1986). Unfortunately, there is little or no information on prenatal growth and development of prosimians except for the birth weights of several species. In fact, except for a few species, information is lacking for the New World monkeys. Most of our growth data is on Old World monkeys, apes, and of course humans.

During fetal life the increase in body weight of the few nonhuman primates that have been studied shows a sigmoid curve ("S" curve) when the extremes of gestation are considered, and most organ growth parallels the increase in body weight. The increase of body segments (e.g., leg length and thigh length) varies in growth rate throughout gestation: one segment may grow faster than the other, or they may change rates during gestation. In the rhesus monkey, the leg grows faster than the thigh and the hindlimb grows faster than the forelimb during the prenatal period.

The importance of differential growth rates in the fetus was also noted in the study by Hendrickx and Houston (1971). They found that baboons and rhesus monkeys grow at about the same rate until approximately the eleventh gestational week, when baboons begin to grow faster. This accelerated growth continues until birth. The result is that the baboon is almost 2 centimeters longer at birth than the rhesus monkey.

The concept of an animal's physiological age is frequently used to estimate progress toward completeness of development. It is based upon the degree of development of different organ systems, which may include bone development (skeletal age), stature or body weight (morphological age), and tooth development (dental age). These systems can be used separately or together to assess the amount of physiological development attained by an animal during both its prenatal and postnatal life. Another use of these physiological ages is the establishment of standards for estimating the gestation age of fetuses of unknown conception date. Skeletal development provides better evidence than do morphological changes, because skeletal stages are less variable and more highly correlated with gestation age.

Investigations of nonhuman primate prenatal skeletal development, like those of body weight and dimensional changes, are scanty at best. The majority of the information comes from laboratory primates, and most of these data are for macaques, although other primate taxa have been studied during the last two decades. Less is known about the prenatal growth and development of prosimians than about those of any other group. This is unfortunate since the prosimians represent such an ecologically and behaviorally diverse group of primates. Skeletal development can be studied by staining the bones with a red dye (alizarin red-S) or by radiographic examination.

Beginning with the early studies of Schultz mentioned above, and supported by all recent work on primate skeletal development, it has been shown that there is a phylogenetic trend to delay the timing of ossification in catarrhine primates. At the same time, their sequences of skeletal development have remained rather similar. Schultz (1972) presented a figure depicting these events in several primate taxa at birth that illustrates the slowdown in ossification quite clearly (fig. 8.6). At birth, Old World monkeys have the most centers of ossification, the apes

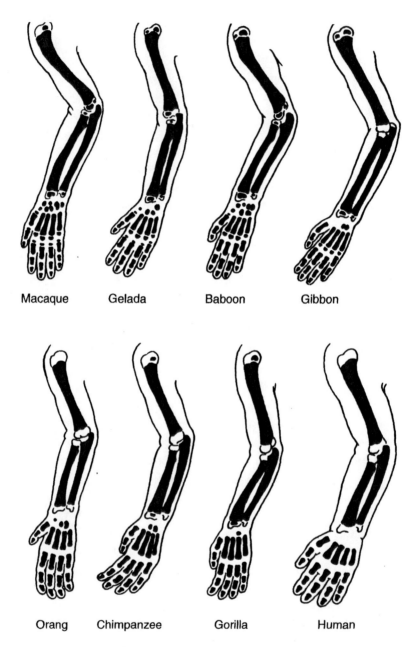

Macaque   Gelada   Baboon   Gibbon

Orang   Chimpanzee   Gorilla   Human

8.6 Ossification of the bones of the forelimbs of several anthropoids at birth. (Redrawn from Schultz 1972)

somewhat fewer, and the human newborn the fewest centers; therefore, the human skeleton is less mature at birth than that of the other primates. Actually, the difference between apes and humans is slight, and it has been suggested that among primates it is the cercopithecines (particularly macaques) that have an accelerated skeletal development at birth (Schultz 1972 and Glaser 1970). For an excellent review of both pre- and postnatal primate skeletal development see Watts (1986).

Another indicator of prenatal maturity mentioned above is dental age. Like skeletal age, dental development can be assessed by alizarin red-S staining of developing teeth or by radiographs. Kraus and Jordan (1965) published what has become a classic study of human dentition before birth. They used alizarin-stained teeth from a large sample of human fetuses to investigate the details of crown formation during gestation. Since then, many studies of deciduous tooth formation in nonhuman primates have been published. Here again there is a paucity of data on prosimians (a recent exception is the study of deciduous tooth formation in the aye-aye [Ankel-Simons 1996]), a little more on New World monkeys, a sporadic representation on the apes, and the most information on Old World monkeys and humans. As with their skeletal development, macaques exhibit precocious development of their deciduous teeth compared with other primates (Swindler and Emel 1990).

An important aspect of prenatal dental development is that there is less variability in dental maturation than in other biological systems, which indicates closer genetic control with fewer outside influences during ontogeny. Dental development is a good indicator of physiological age in primates. In those primates that have been studied, it appears that the cercopithecines are more mature in both their skeletal and dental development at birth than are other primates (Sirianni and Swindler 1985).

It is clear that Old World monkeys are basically precocious developers. They are born after a relatively short gestation period and are relatively independent at about one year. Compared to apes, monkeys begin reproduction very early (about 3 to 4 years) before their skeleton has finished ossifying. Jablonski (1993) has offered an environmental interpretation for this early development which suggests that monkeys became precocious developers as the Old World climates became more seasonal at the end of the Miocene. At this time there was a dramatic turnover from primate faunas dominated by apes to those dominated by monkeys. Apes reproduce successfully only in the tropics, where seasonality is low, where the quality of food is reasonably regular, and where the dependent young do not face a high, weather-related mortality risk in the winter. Monkeys, on the other hand, can reproduce quite successfully from the tropics through the seasonal, montane, high-altitude forests because of their shorter gestations, shorter weaning periods, shorter interbirth intervals, and earlier attainment of sexual maturity. Monkeys, terrestrial and arboreal, follow this pattern, while apes (terrestrial or arboreal) are the opposite.

POSTNATAL PERIOD

There is much more information on the postnatal growth and development of primates than there is for the prenatal period, and this is especially true for the human primate. There are different types of growth data and various modes of analysis used in studying postnatal growth. The methods for collecting growth data are known as longitudinal and cross-sectional studies. In longitudinal studies animals are examined (measured, radiographed, etc.) on more than one occasion when chronological ages (birth dates) are, usually, known. In a cross-sectional study the animals are examined only once since they are placed in an age-graded series usually based on chronological, dental, or skeletal age. All studies of prenatal growth, for example, are cross-sectional.

Both cross-sectional and longitudinal analyses have their merits. Cross-sectional investigations may be done more quickly, but they are somewhat more limited in scope. They are helpful in studying the timing and sequence of developmental events but offer little regarding the rate of growth. They are useful for establishing when particular events occur, like the beginning of tooth formation, the emergence of a particular tooth, or the beginning of ossification of a particular bone. Cross-sectional data offer little information regarding the rate of growth and are of limited value in constructing velocity curves (the study of the *rate* of growth), although they can be used to establish a distance curve of growth, such as height attained at a particular age. The average or mean age of a particular developmental stage such as the emergence of a tooth can be obtained through cross-sectional surveys.

Longitudinal data, on the other hand, are used for establishing velocity curves and for estimating the amount of variability of velocity curves from one period to another. In figure 8.7 notice the obvious adolescent growth spurt in these human velocity curves. Longitudinal data have been widely used in human growth studies (Tanner 1981); cross-sectional data have been used more often in nonhuman primate growth studies.

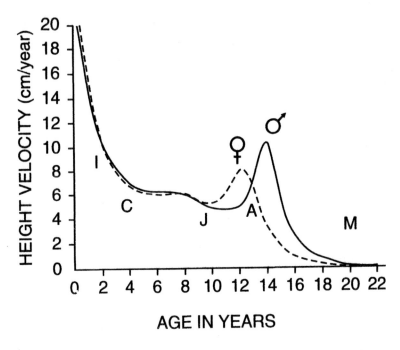

8.7 Human velocity curves for males and females from (I) Infancy; (C) Childhood; (J) Juvenile; (A) Adolescence; to (M) Maturity. (Redrawn from Bogin 1990)

The majority of Adolph Schultz's many publications on nonhuman primate growth and development were cross-sectional, although he published a comprehensive monograph on the growth and development of the chimpanzee in 1940 which included longitudinal information on several animals (Schultz 1940). James Gavan published a longitudinal study of the growth and development of the chimpanzee (Pan troglodytes) which was, and is, one of the most complete analyses of the growth of a nonhuman primate (Gavan 1953). The data represent body measurements taken on 16 chimpanzees born and raised under laboratory conditions. Some major findings were that the chimpanzees required less time to complete their growth than humans and that females took slightly less time than males. In general, chimpanzees grow faster in forearm length, hand length, total forelimb length, head length and breadth; humans have a greater rate in weight gain, hip breadth, upper arm length, and total lower extremity length. The chimpanzee growth period lasts for about 11 or 12 years. The female completes her growth about a year ahead of the male.

In 1956 Van Wagenen and Catchpole published a comprehensive paper on the longitudinal growth and development of a large sample of the common laboratory monkey, *Macaca mulatta*. The data were from birth to approximately seven years, which is the period of postnatal development in these monkeys. Previously growth studies of these monkeys had been limited to very small samples and were mostly cross-sectional. They found that the relatively high rate of growth, as represented by weight and sitting height at birth, slows down during the first six months, levels off at about 18 months of age in females and 24 months in males, then increases in an adolescent "growth spurt" (an accelerated period of growth) which is present in humans and may also be present in other primates (see below).

Another study of the longitudinal growth of macaques was published by Sirianni and Swindler in 1985. The authors report the results of a study of the first eight years of postnatal life in the pigtailed macaque (*Macaca nemestrina*). The period of growth for the pigtailed macaque lasts for approximately 7 or 8 years (similar to that of the rhesus macaque), and, as in most primates, the female is more advanced than the male throughout the growth period. An adolescent growth spurt was reported for these monkeys, and, as in the rhesus monkey, the intensity and duration were greater and longer in the male (Lestrel and Sirianni 1982 and Sirianni et al. 1982). It was found that the pigtailed macaque has a long period of skeletal development when compared with other macaques, and that males, in particular, have a longer postnatal development period than females. It has been suggested that this longer period of bone growth allows for a longer period of linear bone growth resulting in the sexual difference in size in this species (Emel et al. 1982).

The question of an adolescent growth spurt in higher primates has been debated through the years. The controversy is whether the spurt is a uniquely human phenomenon or is also found in other higher primates. Recently Bogin (1988) has suggested that the adolescent growth spurt is a unique feature of the human growth period and does not exist in nonhuman primates. The answer may be a quantitative one, that is, the human growth spurt could simply be large compared to the small one observable in nonhuman primates. This may mean that an adolescent growth spurt is present in all higher primates, the differences among them being simply a matter of degree. Unfortunately there are still only a few long-term studies of laboratory nonhuman primates, and there is almost nothing known of the growth and development of feral nonhuman primates (but see Phillips-Conroy and Jolly 1988).

It is now clear that as we go from monkey to chimpanzee to humans there is a prolongation of the growth period and a decrease in growth rate. For example, Gavan and Swindler (1966) showed that with respect to sitting height (crown to rump), rhesus monkeys reach their maximum size at about 7 years of age, chimpanzees at about 13, and humans not until approximately 18 years. At the same time the relative growth rates change in the opposite direction. The rhesus average rate is approximately twice that of the chimpanzee, which in turn has a rate not quite twice the human average. This trend is perhaps best demonstrated by the classic figure first presented by Schultz (1956) and revised several times (Schultz 1972). The evolutionary trend among primates to prolong their postnatal growth periods is depicted in figure 8.8. The infantile period commences with birth and continues to the emergence of the first permanent teeth. The juvenile period lasts from the beginning of permanent tooth emergence to the emergence of the last permanent tooth. It is during this time that the deciduous teeth are shed and replaced by the permanent teeth and the adolescent growth spurt occurs. By the end of the juvenile period, primates have normally finished their linear growth; we know, however, that one's weight may continue to increase throughout one's life.

The prolongation of immaturity has had its effects on primate behavior. For example, newborn primates are quite dependent upon their parents, especially their mothers (precocial development as discussed previously). It is well known that primates learn by teaching and imitation. The increased length of the learning period before physical maturity enhances the possibilities for learning because the animals have time to observe and mimic the behavior of the adults. Also, prolonged immaturity increases the time that environmental influences can interact with the animal.

The brain is also involved in this evolutionary trend to lengthen the growth period. For example, at birth the brain of the rhesus monkey has achieved approximately 70 percent of its adult size while chimpanzees reach 70 percent of their adult brain size during the first postnatal year. The human brain takes the longest time to reach its adult size; at birth it is about 25 percent, at three years about 70 percent, and by six years it is nearly 95 percent of its adult size. If we allow 20 years for the human postnatal growth period and take 70 years as the average life span, then humans spend 28 percent of their life growing and developing. In a biological sense it takes humans a fairly long time to become adult. Indeed, Bogin (1990) has proposed that only humans have a stage of development known as childhood. Childhood is the time between infancy and the juvenile stage, that is, from about 4 years of age to the beginning of the adolescent growth spurt. This period of delayed

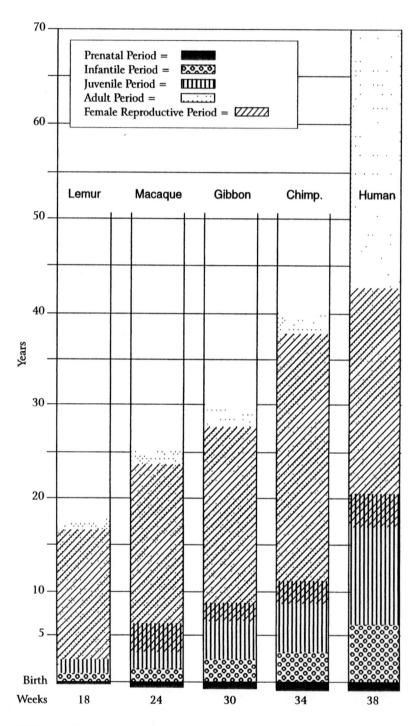

8.8 Primate life spans showing the relative chronogical age of each period within the life of the primate. (Redrawn from Schultz 1972)

maturity permits a time for learning and assimilation of our diverse cultural patterns. As Bogin mentions (1990), the period of childhood is a time during which parents as well as other adult members of the society can interact with children. For example, prolonged childhood, with its attendant learning period, has been associated with the beginning of tool making, food sharing, the establishment of home bases, and perhaps the division of labor. Childhood, then, becomes a most important time in the postnatal development of humans, as was likely the case for many of our fossil ancestors of the genus Homo (Bogin 1990). On the other hand, a recent study of dental development in early hominids found no evidence for delayed maturation and concluded that other, or additional, selective forces may have been responsible for the evolution of delayed maturation and probably postdate the evolution of the australopithecines (Conroy and Kuykendall 1995). Also, studies of chimpanzee dental development support "an interpretation of a rapid, essentially 'apelike' ontogeny among australopithecines" (Anemone et al. 1996, p. 119). The question seems to be when the childhood period made its appearance in the postnatal growth of early hominids rather than whether it is an important factor in development.

In order to compare the differences in postnatal growth between human and nonhuman primates as represented by stature and crown-to-rump lengths, figures 8.9, 8.10, and 8.11 are presented. The distance curve of growth (crown-to-rump in baboons and stature in humans) is depicted in figures 8.9 and 8.10, while a velocity curve (rate of growth as represented by mean length gain) is shown in figures 8.11 and 8.7. The postnatal curve of growth for most animals is an S or sigmoid curve, which indicates that there is an initial acceleration of growth followed by a slower rate of growth which gradually flattens out until adult stature is achieved (fig. 8.10). Notice that there is little sexual dimorphism in stature before the adolescent growth spurt. The human velocity curve (fig. 8.7) shows a marked decrease in growth from birth to about 4 years, after which there is a very slight decrease in growth until about 10 to 12 years in females and 12 to 14 years in males, when the adolescent growth spurt begins. It is this second period, which lasts for about 6 to 8 years, that Bogin (1990) believes is unique in human growth.

The nonhuman primate distance and velocity growth curves are presented in figures 8.9 and 8.11. If we compare the two distance curves in figures 8.11 and 8.7, it is apparent that there is little if any "spurt" of growth along the curve in either sex of the baboon. In general there

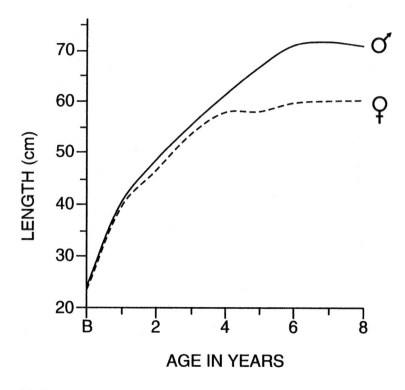

8.9 Distance curves of crown-rump growth in male and female baboons. (Redrawn from Coelho 1985)

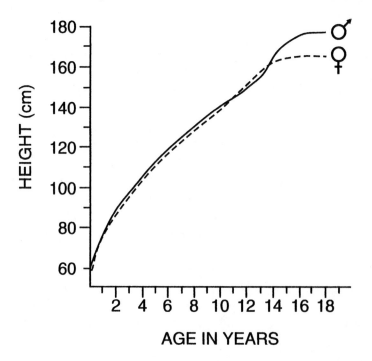

8.10 Distance growth curves of stature for human males and females. (Redrawn from Bogin 1990)

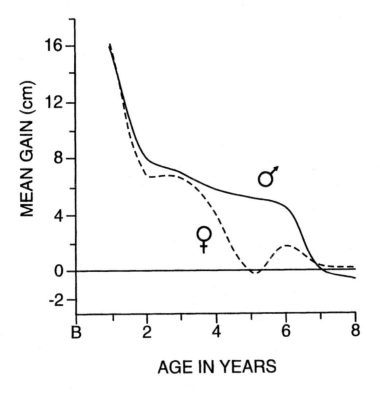

8.11 Velocity curves of crown-rump growth in male and female baboons. (Redrawn from Coelho 1985)

is little sexual dimorphism until about 5 years, when females are approximately 15 percent shorter than males and remain so in subsequent years. Female baboons of this species attain menarche (first menstruation) between 3 and 4 years (Coelho 1985). The velocity curve of mean rate of gain in length is characterized by a gradual decrease in velocity in both sexes. The one major difference coincides with the onset of sexual dimorphism during the fifth year. Again, the major difference between the human and nonhuman primate velocity curves is the marked increase in rate associated with the human growth spurt. In a long-term study of the growth of the pigtailed macaque, Sirianni and Swindler (1985) found growth patterns very similar to those of the baboon.

Long-term studies such as those discussed above represent a collection of data that forms a basis for further comparative studies among primates. It has become clear that many of the differences that exist among primate species are better understood when the processes leading to them are known. Unfortunately, long-term studies require a huge investment, because nonhuman primates are expensive to house and maintain and long gestation and postnatal growth periods are involved. For these and other reasons, the majority of nonhuman primate growth studies have been of the short-term cross-sectional type.

Fig. 8.12 A baboon mother nursing her young infant. Notice the mother's enlarged cheek pouch on the left side of her face. Cheek pouches are discussed in the glossary and on p. 116. (Photo courtesy of Judy Johnson, Regional Primate Research Center, Seattle)

A. Paniscus Goata

9

Social Groups

and Primate

Behavior

Human acquaintance with the monkeys and apes is longstanding, as we have seen in chapter one. Even so, the study of primate behavior is a relatively recent addition to the study of primates. Prior to the twentieth century most information regarding primate behavior was known only from travelers' tales and hearsay, and was mostly anecdotal. It was not until the 1920s and 1930s that it was realized that monkeys and apes were worthy of physiological and psychological study, because it was beginning to be generally accepted that, as our closest structural relatives, they would be valuable in medical research. At about this same time, captive primate colonies were established in several countries (see chap. 1) in order to study primate behavior. Of course zoos had been available for observations for centuries. One of the earlier investigations of the social life of primates was that of Zuckerman (1932) conducted, for the most part, on the hamadryas baboons living in London's Regent's Park Zoo. In 1938 C. R. Carpenter, one of the pioneers of modern field studies, established an early monkey colony on Cayo Santiago Island off the coast of Puerto Rico. It is still available for·research on the rhesus monkey (*Macaca mulatta*). An excellent summary of the history and research done on the monkeys of Cayo Santiago is presented in Rawlins and Kessler (1986). Today, primate behavior is investigated in the laboratory, in artificial colonies, and in the field under natural conditions.

As with most scientific disciplines, the study of primate behavior started as descriptions of the animals and their habitats (Smuts et al. 1987 and Hall and De Vore 1965). Such studies are important because they provide basic information regarding the lives of free-ranging primate populations and serve as a prerequisite for more advanced, problem-oriented studies. Only a little over thirty years ago, Southwick (1963, p. 5) was able to write, "For most species of primates, our field knowledge is surprisingly sparse or non-existent." Since then, a wide variety of field studies have been undertaken by a cadre of scholars representing such disciplines as anthropology, zoology, and psychology. The effect has been a tremendous increase in interest and concern for wild living primates, which has resulted in a burgeoning literature on primate behavior during the past three decades. It is clear that primates display a wide range of behaviors and social organizations that vary from species to species as well as within species living under different environmental situations such as different predation intensities and different population densities. It is, therefore, not possible to present the multidimensional aspects of primate behavior in this one chapter; rather, a general statement of selected aspects of primate behavior is presented from contemporary field studies.

Social Behavior

As discussed in the last chapter, primates have a long postnatal growth period, which offers the developing primate a protracted period for learning, and which in turn permits "the development of more flexible, less programmed, behavioral responses to different situations" (Napier and Napier 1985, p. 61). Primates pass through the following developmental stages: 1) infant, a time during which a primate is dependent on its mother; 2) juvenile, during which it is independent of its mother but not sexually mature; and 3) adult, during which it is a mature animal. Recently Bogin (1990) has stated that the juvenile stage is unique to primates and some of the social carnivores, but that only childhood, which is inserted between infancy and juvenescence, is unique to humans (see chap. 8). The primate juvenile period is a time of great learning, not only from parents, but also from interaction with other members of an animal's immediate group (Pereira and Fairbanks 1993). Many of the social skills that are necessary for life in the group are learned during this period. It is during early adulthood that the secondary sexual characters begin to develop, for example, the long, robust male canines and the increase in male body size compared to females, which play important roles in the behavior of those species possessing sexual dimorphism. This extended period of growth and development is one of the hallmarks of the order Primates.

Primates are gregarious, social animals who travel, eat, and live together in groups, although there are a few species that are more solitary than others, such as orangutans, tarsiers, and galagos. Primates possess many different ways of expressing themselves socially, such as touching, hugging, mounting, grooming, vocalizing, and lip-smacking, which allows a far richer repertoire of social behaviors than are found in most other animal species. Primate social organization is complex since there are different sexes, ages, kinship, and dominance ranks represented in the groups. In addition, these relationships are frequently cross-cut by subgroups composed of short- and long-term alliances, which makes for increased complexity. For example, grooming alliances may not have only short-term consequences; long-term relationships are also established when males groom a pregnant or lactating female, an activity which may enhance receptivity by the female when she resumes sexual cycling. Another factor running through most primate societies is that many of these alliances shift, and individuals change through time, so that the primate group is a dynamic unit as well. It should be understood, however, that current research clearly demonstrates that many social relationships are much less uniform across primate species than originally thought, thereby "raising questions about the generality of models of primate social systems derived from 'typical' primates" (Strier 1994, p. 233).

There are really no solitary primates. Even the so-called solitary species (e.g., orangutans) are at best semi-solitary since the males and females come together to mate and the female and one or two infants form social groups. An adult male orangutan is usually in the neighborhood of one or more of the female social groups. They are certainly the least gregarious of all diurnal primates.

The areas in which primates carry out their daily or nightly routines of living are defined by the activities normally carried out in them. For example, a home range is an area occupied by a primate during its adult life. It is the region where the group lives. It is not defended and may overlap with the home ranges of other groups. The sizes of home ranges vary considerably for a variety of ecological and social reasons, among which are the size and proximity of neighboring troops and the concentration of food plants. In figure 9.1 the home ranges of several primate species are shown. It is apparent that the more arboreal species, the gibbons and howler monkeys, have somewhat smaller groups and tend to have more limited areas. The langurs, on the other hand, live and feed in both the trees and on the ground and use larger areas. Of the different langurs, the hanuman langur is the most terrestrial. The largest areas are occupied by the more terrestrial species, the baboons and the gorillas.

Within a home range is a more restricted area known as the core area. The core area includes the food trees used by the animals and their sleeping trees, and it is the area that is most used by the group. Figure 9.2 depicts the home ranges and core areas of nine troops of baboons living in the Nairobi Park in Kenya. It is obvious that there is considerable overlap of home ranges but very little overlap among the core areas. A third region, the territory, is an area that is occupied and used exclusively by a group who will defend it against intruders. Territories that overlap are a constant source of trouble and can lead to outright aggressive behavior. Aggression, however, is variable between groups and is also present in nonterritorial species. When primate species occupy the same geographical regions, whether on the ground or high in a forest canopy, they are referred to as sympatric species. When species do not have overlapping areas they are known as allopatric species.

In general, primates are conservative animals that live rather monotonous lives. Their lives are generally spent within the home range and with the same social group. Some may join new groups, but under normal conditions this does not occur very often. Tree-living primates usually live in smaller groups and wander less widely than terrestrial species. It seems clear that gibbons and howler monkeys are able to find sufficient food within rather limited areas. Hanuman langurs often feed on the ground, and use wider ranges like the more terrestrial baboons and mountain gorillas. It is generally true that the more arboreal species have not only less vertical but also less horizontal travel during their

GIBBON
1/10 sq. mi. range for group of 4

HOWLER MONKEY
1/2 sq. mi. range for group of 17

NORTH INDIAN LANGUR
3 sq. mi. maximum range for group of 25

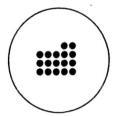

MOUNTAIN GORILLA
10-15 sq. mi. range for group of 17

BABOON
15 sq. mi. range for group of 40

9.1 Home ranges of some tree- and ground-living primates. Circles represent the home ranges, black dots the number of individuals in the group. (After Eimerl and De Vore 1977)

daily wanderings. All guenons are basically tree-living and occupy different levels of the canopy. One species, *Cercopithecus aethiops* spends considerable time on the ground foraging for food while others spend virtually their whole lives in the trees. The colobus monkey is arboreal and essentially folivorous, usually preferring the middle to higher lev-

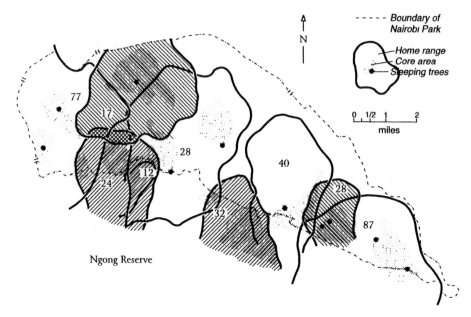

9.2 Home ranges and core areas for nine groups of baboons. The numbers on the map refer to the group size. (After De Vore and Hall 1965)

els of the forest and rarely leaving the trees. Also, fruit-eating species have larger ranges than species of similar body weight who feed on leaves.

The composition of primate groups differs greatly among the species, the result of the operation of many different selective forces. Several types of groups have been defined, but it is clear today that primate social groups are much more dynamic and fluid than originally suspected. This is due in part to the fact that most of the earlier field studies undertaken were of short duration, perhaps several weeks to several months, and long-term information was not available for analysis until relatively recently. A short discussion of the major primate social groups follows.

What is usually considered to be the most primitive social group among primates is the noyau group, which consists of an individual female and her offspring (Charles-Dominique 1983). Although the noyau arrangement is found among some nocturnal prosimians (e.g., the African bush babies), it is also the type of social organization practiced by the diurnal orangutans (Fleagle 1988). The adult males use much larger areas than those of the females in their daily foraging for food, and their travels may cross the core areas of several of the female groups, undoubtedly helping the male to monitor the whereabouts of the females.

The multi-male group consists of several adult males and several adult females with their offspring. This pattern of social organization is found in many species of primates, including prosimians (the ring-tailed lemur), platyrrhines (the mantled howler monkey), Old World monkeys (the rhesus monkey, olive baboon, and red colobus monkey), and the common chimpanzees, to mention a few. The presence of more than one adult male in a group frequently leads to tension and friction among them and can result in the formation of a dominance hierarchy among the males of the group. In this system there is only one dominant male at a time (see below). It should be noted that dominance hierarchies are more common in cercopithecines than in colobines or in most New World monkey species. The dominance hierarchy is also present among females. The result is a system that seems to work fairly well for both sexes. The adult male is generally dominant over the female, particularly in those species with marked sexual dimorphism.

One-male groups have a single adult male with several adult females and their offspring. This structure is found in the prosimian sportive lemur (*Lepilemur mustelinus*), in some Platyrrhini such as the spider monkey (*Ateles geoffroyi*) and the howler monkey (*Alouatta seniculus*), in several terrestrial Old World monkeys such as the patas monkey (*Erythrocebus patas*), as well as in several colobine monkeys belonging to the genera *Colobus* and *Presbytis*. Among great apes, the gorilla has one-male groups, although gorilla social organization is more often characterized by what is known as age-graded groups (see below).

An intermediate type of social structure between the groups described above has been identified and described as the age-graded male group (Eisenberg et al. 1972). The major difference between this system and the multi-male system is "that only one male does most of the breeding, and he is thought to be related to the other, usually younger, males" (Crockett and Eisenberg 1987, p. 65). A good example is the gorilla, which lives in groups of ten to fifteen animals that consist of only one dominant, older adult male (a silverback), several younger males, and females with their offspring. The silverback male is clearly the leader and social focus of the group. He governs the time and direction of group travel, settles intragroup quarrels, and affords protection for the group. When more than one silverback is present in a group, the oldest one is the leader and the others occupy peripheral positions. As the males get older and become silverbacks they may move away from their group. They are then frequently joined by nulliparous females from their own or other groups, which results in the formation of new groups. As these new groups form, the silverback quickly becomes the leader of its group.

The family group is similar to the human nuclear family and consists of a bonded pair of adults and their immature offspring. This is not a common type of grouping among nonhuman primates, and it was first described by Carpenter in 1940 in the gibbon, *Hylobates lar*. Since then this form of social organization has been described in the lemur, *Indri indri*, the night monkey, *Aotus*, and the titi monkey, *Callicebus moloch*. The primates living in these monogamous groups appear to stay together for life; when their offspring reach sexual maturity they leave or are run off by the adults and form their own families. Among titi monkeys the father also spends much time in the care of the offspring, actually spending more time in contact with the young than does its mother or its siblings. Another important role of the father is to carry the young animal as the group moves through the trees. This behavior also contributes to the breeding success of the male, because he is protecting his own genetic legacy to the next generation.

It is obvious that primates are social, group-living animals that have evolved a mixture of complex social organizations through the process of natural selection. At one time it was thought that the central reason for group living in primates was the sexual attraction of males for females, since it was believed that female primates had estrous cycles throughout the year (Zuckerman 1932). It is now known that many species of primates, from prosimians to Old World monkeys, have restricted breeding periods that result in births being limited to certain seasons of the year, often when there are abundant food sources available. It is apparent today that more factors than sexual attraction are required in order to explain the multifactorial nature of primate social organization. Other important benefits of group living are increased protection from predators, a wider access to mates, better reproductive success and assistance in raising offspring, and better access to food sources. These factors are both biological and environmental, and all have played important and vital roles throughout primate history. If there is one clear message coming out of recent studies of primate behavior it is that "we have hardly begun to grasp the richness and complexity of these animals' social lives" (Smuts et al. 1987, p. 3).

Because most primates live in groups, the animals must compete with one another from time to time for food and mates. This competition results in the establishment of a hierarchy of dominance that involves individual differences in behavior based on sex, size, age, status, and kinship. Dominance lets an animal know where it is in the group and what it can and cannot do with respect to other animals, thereby reducing many causes of intragroup agonistic behavior. Thus, dominance is adaptive in that it operates to lessen conflict in the group.

The establishment of dominance frequently depends on size and strength, but certainly not always. The most obvious manifestation of competition within a group is aggressive behavior in one of many possible forms. There are many cases where older male animals maintain their dominant status simply by acting like a dominant animal. It is true that males are usually dominant over females, but here again there are many exceptions where females are dominant over males, or at least codominant.

Smuts (1987) describes various forms of dominance in primate species.

1. Species with little body-size sexual dimorphism or in which the females are somewhat larger than the males. The females are dominant in these species which include many of the lemuriforms.
2. Species which have slight differences in body size and in which the sexes are codominant. When agonistic interactions occur between the sexes, generally neither sex dominates the other. This behavior pattern occurs in callitrichids, monogamous platyrrhines, and monogamous gibbons.
3. Species where the males and females are of similar body size but the males dominate the females by male-male coalitions. The only known species possessing this arrangement is the spider monkey.
4. Species in which the males are larger than the females but the females may dominate the males by female-female coalitions. These groups include squirrel monkeys, vervets, bonobos, and perhaps the patas monkey.
5. Species in which the males are larger than the females and the latter almost never dominate the males. These species include gelada, hamadryas, and savanna baboons, the great apes, mantled howlers, and gray langurs. There are still many species that cannot

be classified because there is not sufficient information regarding their social organization. This is especially true for the nocturnal prosimians.

The dominance hierarchy is often referred to as a "pecking order," something originally described among birds. This type of behavior is particularly well developed in savanna baboons and rhesus monkeys. Indeed, multi-male groups with their dominance hierarchies were first identified in the baboon, *Papio anubis*, in the pioneer study of DeVore and Washburn (1963). Many of the dominance hierarchies are linear, with an "alpha" male at the top, followed by a "beta" male, who in turn is followed by a "gamma" male, and so forth. On occasion, the "beta" and "gamma" males link up to run off the "alpha" male, and then the "beta" male assumes leadership of the group if he maintains his dominancy over the "gamma" male. Such changes in power can be disruptive to the group, causing more tensions and grievances among the members of the troop until the new "alpha" male takes over. It should be understood that the dominance hierarchy is generally not really linear, since an individual may be dominant in one situation but not in another.

Female dominance is not common in primates, or in mammals generally. As we have seen, it is present among lemuriforms, but females are also sometimes dominant in *Cercopithecus talapoin*, some macaques, and *Cebus apella*. In the latter forms, females are only dominant over males when they form coalitions with other females. In some species a pregnant female that is subordinate when not pregnant may be dominant to males. It is unlikely that female dominance in primates is homologous, that is, derived from a common ancestor. There is not a lot of information available, but when only the lemuriforms are considered, female dominance appears "to be represented by a distinctive and homogeneous set of behaviors, notably the consistent displacement of males by females at feeding sites" (Richard 1987, p.32).

Mother-offspring relationships are also important since, in many species, the status of the mother is transferred to her offspring whether she is a high-ranking or low-ranking female. Thus, in the Japanese macaques the offspring of high-ranking mothers enjoy a more dominant position than those of lower ranks. In addition, male offspring are ranked below female offspring, which may indicate the separation of gender roles (Napier and Napier 1985).

Grooming

Grooming is a tactile activity practiced by all primates from prosimians to great apes. It is a type of social interaction in which one animal is groomed by another. In other words, there is a groomer and a groomee, and from time to time the roles may be reversed. Grooming is usually initiated when one animal (the groomee) approaches another member of the group and presents some part of its body for grooming, whereupon the second animal (the groomer) begins inspecting its skin and picking off pieces of dirt, flakes of skin, and any ectoparasites that might be present. Grooming may well have started as a cleaning or hygienic behavior which in time developed into a much more important socializing behavior since it helps in reinforcing social relationships and attaining proximity to more dominant animals.

It is well known that grooming takes place more often among related than among nonrelated individuals in many primate species, and in many cases it is among mothers and their offspring (Gouzoules and Gouzoules 1987). However, grooming does occur among nonrelated individuals in many primate species, even in those that groom mainly kin. In most groups infants usually receive more grooming than older offspring.

Prosimians groom somewhat differently from other primates in that they use their tooth combs (lower incisors and canines) and toilet or grooming claws (on the second digit of the foot) for both self-grooming and the grooming of other members of their species. However, they seem to use their dental combs more often when grooming other individuals and generally reserve their grooming claws for cleaning and maintaining their own fur.

Grooming is present in all great apes. Among gorillas grooming takes place generally between silverback males and parous females, although the frequency and the direction of grooming differ widely among gorilla groups (Stewart and Harcourt 1987). Grooming in orangutans has been observed between males and their consorts as well as between females when they are together. Chimpanzees groom a great deal (fig. 9.3), particularly mothers and their offspring, an activity that certainly helps to strengthen family relationships. Male chimpanzees also spend long sessions grooming each other.

It is obvious from the above discussion that grooming among primates serves many functions, from hygiene to preserving group stabil-

9.3 Bonobos grooming. (Photo courtesy of Ellen Ingmanson)

ity and solidarity. Because they reduce tensions, reinforce family ties, and maintain dominance relationships, grooming sessions are an important and vital force in maintaining the cohesion of the primate group.

Communication

Communication is especially important to all social animals, such as the primates, who live in groups. They have evolved various ways to communicate with each other; consequently, some type of communication pervades all aspects of their lives. Some use scent (olfactory) signals, some employ visuals aids, while others use auditory signals.

Olfactory communication is generally used by primates for sexual interactions, although it is used in other connections as well. Other types of communication such as morphological and visual signals are used to attract males. In many species the males smell the anoperineal region of the female or even examine her vagina in order to monitor her reproductive status. Among prosimians, olfactory signals attract males to the females. At the height of estrus there may be several males following one female. Special scent glands in various areas of the body (e. g., in the armpit, along the forearm, or near the base of the tail) secrete pheromones whose odors attract and excite males (see chap. 6). Primates may distribute scents by scent-marking on branches of trees or by urinating on their own hands and feet to indicate their trail as they walk or climb through the forest.

Pheromones are used for other than sexual attraction. Many primates use scent markings to establish territories, as alarm mechanisms, and in aggressive encounters. One interesting display that is both olfactory and visual is the so-called stink-fight of the ring-tailed lemur. The stink fight is frequently used by males to establish territorial boundaries or in competitions over estrous females. In these displays, the males rub their chests and forearm scent glands together, stimulating the flow of pheromones, and then rub their tails through their forearms and across their chests, after which they quiver and shake their tails at their adversary, who receives an odiferous blast from the waving tail. The result is that either one or the other will retreat after several of these bouts, usually without physical contact.

Of course willingness to mate can be indicated in other ways, such as gestures, facial expressions, and different types of postures. Among some lemurs and marmosets crouching and bending forward is employed, while in tamarins and howler monkeys flicking the tongue is used to attract males as is puckering the lips in capuchins and patas monkeys (Hrdy and Whitten 1987). It should also be mentioned that males of many species, such as baboons and rhesus macaques, solicit females to mate by lip smacking and various facial expressions.

Visual signals make up an important system of communication among primates. Indeed, the enhanced visual communication among diurnal primates has, in a sense, replaced the varied olfactory signals of the nocturnal prosimians. It has become, along with vocalization, one of the two dominant forms of communication among the anthropoids.

Visual signals include body postures, gestures, facial expressions, and the different positions of the tail. The latter is particularly meaningful in rhesus monkeys, where the tail of the dominant male is held high above its body with a slight S-curve near the end, while a lower-ranked male carries his tail low below the level of his back while walking (fig. 9.4). In recent years it has become apparent that many aggressive and friendly gestures share similar components, which suggests that the components act collectively in a particular gesture to mean a certain action. It has been found in Barbary macaques, for example, that staring, lowering and raising the eyelids, and the erection of hairs (pilo-erection) are the four most common components in aggressive behavior

9.4 Macaques display dominance with their tails. Animal above has its tail raised and is dominant over the animal below with its tail lowered.

while, in a complement of seven friendly gestures, they also make up four of the seven (Zeller 1987).

It is clear that among nonhuman primates visual signals are used in both friendly and competitive interactions to manipulate various social relationships. These ritualized visual behaviors tend to lessen tension within a group by regulating and maintaining group cohesion by preventing physical fights. For example, at any time during a series of displays by a dominant male macaque, the subordinate animal can terminate the action by performing a conciliatory gesture such as presenting its rear end to the dominant male who, in turn, may mount for a short period. This type of behavior is common in many Old World primates, and Napier and Napier (1985, p. 75) have suggested that it is "similar to and possibly derived from the mating postures of male and female."

The different facial expressions of animals, especially those of primates, have long interested zoologists. In fact, it was Charles Darwin who first called attention to the muscles of facial expression in humans and other animals and proffered explanations for the underlying emotions associated with various expressions. That the yawn in many primates is often a threat signal, particularly when the upper canines are exposed as in baboons, was first recognized and discussed by Darwin (1872). The yawn, as with many facial expressions, may convey several meanings depending on the circumstances and the intensity of the expression. In many monkeys the invitation to approach is a lip-smacking gesture often accompanied by tongue thrusting. The play-face, or grin, is an open mouth expression in which an animal looks toward but does not stare directly at another animal. The play-face is a friendly gesture, but if it is done while staring or glaring directly at another animal it is a threat gesture in most primates, including humans.

Auditory communication is common in most primate species, especially among diurnal forms. Many arboreal species have warning calls to alert the members of a group of impending danger from a bird or snake. Vervet monkeys have special vocalizations for indicating snakes, leopards, and raptorial birds (Cheney and Wrangham 1987). Chimpanzees have various vocalizations such as the pant-hoots which they make during feeding, when groups meet, or when they begin to nest in the evening. These calls also help to identify the caller.

In the early morning, and usually in the evening, in forests from northern Argentina to Mexico there can be heard a loud, deeply pitched roar that resonates through the otherwise quiet forests. This vocalization is made by the howler monkeys, noted for their response to the vocalizations of their associates (fig. 9.5). According to Carpenter's classic paper (1934) on the naturalistic behavior of howler monkeys (Alouatta palliata), there are probably fifteen to twenty separate vocal patterns in their repertoire. They include deep, hoarse clucks that initiate troop movements; gurgling sounds made only by males as defensive vocalizations; and the wailing sounds made by a mother when her infant has fallen or otherwise been removed from her.

Another primate famous for its vocalization is the gibbon. As with the howler monkey, gibbons begin their calls in the early morning as the monogamous families break up from the groups in which they have spent the night and begin their daily routines. The calls stimulate those from adjacent groups, and before long the crescendo rises into what is called the "great call." These stereotyped calls are actually territorial songs that serve to identify the boundaries among gibbon territories and are unique for each species (species-specific). During the day as gibbons travel and eat they will vocalize as they get near to another group which responds with its call. The two groups will engage in a vocal battle that becomes louder and faster until one group moves away, thus lessening the chances of actual physical contact. Extensive studies

9.5 Red howler monkeys (*Alouatta seneculus*) howling at a neighboring troop during a confrontation in an area where their home ranges overlap. Adult females at left and far center; adult male right center; large juvenile males far right and top (head not seen). (Photo courtesy of Carolyn M. Crockett)

of gibbon vocalizations by Marshall and Sugardjito have led them to state that gibbon songs are inherited and that therefore "the territorial song is a useful criterion of the species in gibbon systematics" (1986, p. 177).

A frequently asked question is, "do apes (especially, the chimpanzee) have language?" This is a difficult question to answer. Certainly the three great apes have been taught to use sign language (American Sign Language), and particularly the chimps have used it to communicate with each other, and have even taught other members of the group to use signs. In this sense nonhuman primates communicate with each other, but is that language? Human language with its grammar and syntax is usually considered a unique possession of Homo sapiens. A lot depends on how language is defined as well as on the scholars making the definitions. This is an interesting but complex subject which warrants a more intensive discussion than can be done in this chapter. Here we can only briefly review the subject of language and the great apes.

Earlier investigations attempted to teach apes the human spoken language (Hayes 1951). It was later realized that apes lacked the proper laryngeal anatomy for articulating human words. Moreover, it was demonstrated by Duchin (1990) that the anatomical relations of the oral cavity in chimpanzees prohibit the tongue from making the articulate sounds of human speech. When investigators began to use sign language instead of spoken words, a whole new area of research was opened up. A chimpanzee by the name of Washoe, raised by the Gardners in their home, was taught more than a hundred signs in little over three years (Gardner and Gardner 1969). Today, Washoe is at Central Washington University and is under the supervision of Roger Fouts. What is most interesting, and rather unexpected, is that Washoe has taught signs to another chimpanzee (Gardner and Gardner 1969). Chimpanzees have also been taught to use drawings as symbols for words: touching a certain figure, for example, indicates a particular word. There are, of course, many questions to explore in this research, but certainly much more progress has been made using sign language than was achieved in the earlier studies that attempted to use human vocal expressions.

It is apparent that nonhuman primates do possess vocal calls and gestures representing different objects (e.g., snakes, birds, and food sources), and this can reveal how the animals perceive and classify things in their natural environment. Such information has allowed Seyfarth (1987, p. 451) to write that primate "vocalizations are thus not only of interest in themselves, but also provide a tool for studying social structure from the animal's point of view." Our concept of primate communication has changed a great deal over the last couple of decades, and the more we learn about nonhuman primate communication systems the less distinct the line becomes between nonhuman primate and human communication.

Tools

Until rather recently it was believed that, of all creatures, only humans made and used tools. Many animals were known to use unmodified objects as tools (Beck 1980), but fabricating tools was the unique activity of the genus Homo. Then, in the 1960s, Jane Goodall reported that chimpanzees made and used tools as probes for what has become known as termite fishing (Goodall 1970). The chimpanzee begins by making a hole on the surface of a termite mound into which it pushes a blade of grass, a piece of bark, or a long twig. These tools have usually been modified by the chimpanzee on the way to the termite fields by stripping off the leaves or breaking the twigs into different shapes for later use at the termite mounds. When the stick is inserted into a hole, the termites attach to it, and when the stick is removed, the chimpanzee licks off the termites. There is also a sex difference in the frequency of termite fishing among the Gombe Stream chimpanzees. It seems that females spend much more time termite fishing than do males (McGrew 1979).

The important feature about chimpanzee tool use is that they prepare the tool for use before the event; indeed, they tailor their tools for a particular job, and different chimpanzee groups use different types of tools. They have also been observed making more than one tool at a time and carrying them some distance to the termite mounds. The repertoire of chimpanzee cultural behaviors includes, in addition to termite fishing, nest building, hunting, and throwing sticks and stones as offensive weapons.

Tool use has been reported occurring occasionally among the wild orangutans living in the Tanjung Puting National Park of Borneo (Galdikas 1982). Such uses include waving sticks to ward off wasps or to scratch one's back. Captive orangutans are famous, or perhaps infamous, as lock-picking escape artists. Zimmer (1995) reports that Van Schiak, a primatologist from Duke University studying orangutans in the Suaq Swamp of Sumatra, has observed them using prepared branches to stick in holes to obtain termites or honey. It appears that the gorilla is the only great ape that does not use tools in its natural habitat.

What may be called cultural behaviors are also found among other nonhuman primates. Monkeys, for example, have been observed cracking nuts against trees or rocks to obtain the kernels. The Japanese macaques (*Macaca fuscata*) living on Koshima Island are almost legendary for their cognitive behavior. The Japanese researchers supplied the monkeys with sweet potatoes for food and found that one young female began a practice of washing the sweet potatoes in the nearby sea. The habit was quickly picked up by other members of the group. The sweet potatoes were not only cleaned by the washing but the monkeys preferred the added taste of salt to the taste of those washed in fresh water. But neither the Japanese macaque nor, for that matter, any other monkey, has ever been observed making tools.

The last two decades have witnessed an increase in the collection of information on primate behavior that has resulted in the testing of many hypotheses regarding the evolution of primate societies (Smuts et al. 1987 and Strier 1994). This adds immeasurably to a deeper understanding of the complexities of primate social organization, which, in turn, may help us to better understand our own human behavior.

BEHAVIOR AND ANATOMY

It is apparent that primates have many different types of social organizations with different sex ratios, different reproductive strategies, and different age grades. It is also obvious that there are numerous anatomical and physiological characteristics associated with these various social groups. Canine dimorphism in nonhuman primates is related to several selective pressures such as sexual competition among males for females, predator defense, and sexual differences in body size. Canine size is associated with certain aspects of each of these: males that maintain harems, compete among themselves, or are terrestrial have larger canines than do monogamous or arboreal primates (Harvey, Kavanagh, and Clutton-Brock 1978). In general, males and females of monogamous primates (gibbons) have similar body and canine sizes. Plavcan and van Schaik (1992) also demonstrated the strong degree of canine dimorphism in social groups where intermale competition is present. Predation and body-weight dimorphism were also associated with canine dimorphism, although not to the same degree as intermale competition. In a study of dental arch form in cercopithecids, Siebert and Swindler (1984) found that species grouped by predator defense showed more sexual dimorphism in their dental arches than those grouped by diet or social structure.

The length of limbs determines, to some extent, the kinds of food a primate can reach. Many folivorous primates prefer the young, tender, and more nutritious leaves that are often out on the ends of branches, and longer limbs help them stretch to these leaves. In like fashion, the long, prehensile tails of howler monkeys are helpful when they use them to swing outward from a branch to reach the more succulent leaves at the end of other branches. In many cases, primate species can live sympatrically in the same forests by exploiting different food sources due to differences in their anatomy or physiology. For example, the strong jaws and teeth of mangabeys, particularly the large upper incisors with their thick cover of enamel, permit them to bite into the tough coverings of many fruits to obtain their contents, while colobus monkeys eat leaves in the same forest and slowly process them in their large, sacculated digestive systems with their great quantities of microbes.

Truly, behavior is everything that an animal does, not just its social behavior. It is worth emphasizing again that knowledge of the relationship between anatomy and behavior in extant primates, especially the correlates between dental morphology (including microwear patterns) and diet, has proven extremely helpful in interpreting the behavior of fossil primates (Teaford 1994).

10 ~

Fossil

Primates

CENOZOIC ERA, 65–2 MILLION YEARS AGO

A geological time scale for the last 65 million years of earth history is presented in table 10.1. This period of time is known as the Cenozoic era and is often referred to as the Age of Mammals. During the Cenozoic era primates appeared and adaptively radiated into all the various species living today. It is the emergence, diversification, and geographical distribution of the primates during this long time interval that will be considered in this chapter. It is necessary, however, to take a quick look at the paleoclimate and geological events taking place during this time in order to understand the conditions in which the primates evolved.

At the beginning of the Mesozoic era, some two hundred million years ago, all the major continents of the earth were locked together in one supercontinent known as Pangea, "all earth" (fig. 10.1). Some time

Table 10.1   Cenozoic Time Table. Time (MYA*)

Era	Period	System	Epoch
Cenozoic (65 MYA to present)	Quaternary (1.8 MYA to present)	Quaternary (1.8 MYA to present)	Holocene
			12,000 yr
			Pleistocene 1.8 MYA
		Neogene (24–1.8 MYA)	Pliocene 5 MYA
			Miocene 24 MYA
	Tertiary (65–1.8 MYA)		Oligocene 35 MYA
		Paleogene (65–24 MYA)	Eocene 55 MYA
			Paleocene 65 MYA

Million Years Ago

later, Pangea began to separate into various land masses and to move in different directions, ultimately forming the continents we know today. The phenomenon of continents slowly drifting around on the surface of the earth is now an accepted geological fact and referred to as plate tectonics. Initially Pangea was separated by a large body of water, the Tethys Sea, which split the supercontinent into a northern land mass, Laurasia, and a southern continent, Gondwanaland. Laurasia then broke apart to form North America, Asia, and Europe, while Gondwanaland ruptured into Africa, South America, Antarctica, Madagascar, Australia, and India. When continents move, new ocean floors form by the

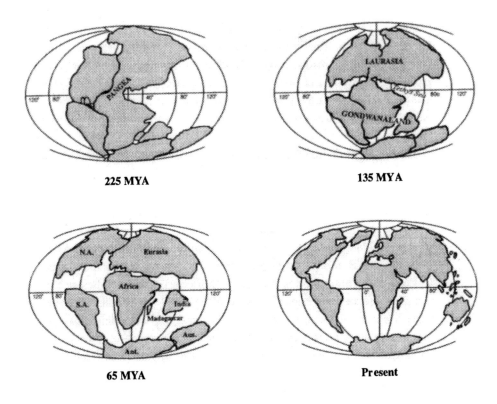

225 MYA

135 MYA

65 MYA

Present

10.1 The restless earth, or plate tectonics, over the past 225 million years.

growth of the plates on which they ride. The Atlantic Ocean, for instance, is still growing at a rate of from 5 to 7 centimeters per year as the American continent continues to draw away from Africa and Europe.

The aspect of plate tectonics that is important to the study of animal and plant history is paleogeography, or the plotting of the courses of continental migration during their long journeys across the surface of the earth. Where continents were at different times over the last two hundred million years or so has helped to explain many puzzles regarding the past and present distribution of animals and plants. For example, the small lizardlike fossil reptile Mesosaurus lived during the late Permian era, some 250 MYA. Their remains have been found only in eastern South America and southwestern Africa (Raymo 1983). The explanation for this apparently unusual geographic distribution is that Mesosaurus lived before Pangea separated, and its remains were carried on the two continents as they moved to their present locations. Another example helps to explain the similarities among some of the land-mammal fauna in North America and Europe during the Paleocene-Eocene transition. During this time the climate was warming, and many taxa, including primates, migrated northeast out of North America to cross high-latitude land bridges between the two continents (Gingerich 1989). At the same time portions of northwestern North America were connected to northeastern Asia in the region of what is today the Bering Strait.

As continents drifted they usually created water barriers which slowed down or completely halted animal migrations. Continents, however, could not continue to move apart without suffering new encounters with other land masses. One of the more outstanding collisions occurred at the end of India's long journey across the Tethys Sea some 45 MYA (middle Eocene) when the subcontinent collided with Asia. This culminated in the world's greatest mountain system, the Himalayas. It was several million years later (early Miocene), however, before the thrusting Indian plate began the vertical uplift of the Himalayas. Since India has not yet come to rest, the Himalayas are still rising three or four centimeters a year. Eventually erosion will outpace uplift and the Himalayas will be "cut down to size" like the Appalachian mountains along the east coast of North America. Such old ranges indicate the junctions of earlier continental collisions (Raymo 1983).

With the rise of the Himalaya mountains there was a contraction of the Tethys Sea. One arm of this ancient body of water remains today as the Mediterranean Sea. These events resulted in a gradual change from a warm, humid climate in southern Eurasia and northern Africa with closed forests to a drier, cooler climate with a more open woodland forest (Laporte and Zihlman 1983). We will see the impact that some of these environmental changes had on primate evolution later in this chapter.

These few examples from paleogeography should indicate the important role of environment on biological evolution. As the continents moved across the face of the earth, mountain systems were formed and subsequently eroded, affecting major water sheds and wind directions. Paleoclimates changed (there has been a general world cooling trend since the late Eocene). Forests changed into savannas, then into arid deserts, and back again into forests. Sea levels changed drastically through time, when low land bridges were formed between adjacent land masses for various periods only to disappear as sea levels rose. The Cenozoic era, some 65 million years in duration, saw many of these changes. There is little doubt that the generalized cooling of the earth's temperature, contraction of the large tropical forests, advent of the extensive grasslands, and orogenic (mountain building) uplifts played vital roles in establishing new habitats for primate explorations. This was a time of great maelstroms both geological and biological, and it was to witness the origin and diversification of the primates, including the origin of our own species, Homo sapiens.

PALEOCENE EPOCH, 65 – 55 MYA

The Paleocene was generally warm, with savannas and woodlands dominating the landscapes in North America and Europe. There were no seasons as there are today, thus no extreme temperatures, and North America was still attached to both Europe and Asia. Although they had been present for millions of years, the early mammals were very small and probably arboreal. A recent mammalian skeleton from the Jurassic tends to support the arboreal theory for the origin of mammals (Martin 1993). Indeed, it is now well known that there was considerable diversification of mammalian lineages before the end of the Mesozoic, so that by the end of the Cretaceous mammals were beginning to stir and make their presence known. After all, dinosaurs were now extinct and the great mammalian revolution was about to begin. The Age of Mammals had arrived.

Primates, it has been generally agreed in recent years, appeared from a basic insectivorous mammal, not unlike modern hedgehogs or treeshrews, probably during the Cretaceous-Paleocene transition. There is even a fossil candidate in the form of a small, mouse-sized creature known as Purgatorius from late Cretaceous-Paleocene beds in Montana (Van Valen and Sloan 1965). The original identification was based on several isolated teeth (there was no skeletal material); the cusps of the molars were sharp and relatively high, suggesting that they ate insects, as many prosimians do today. Later several mandibular fragments with teeth were found in the same site and the structure of these teeth supported the original diagnosis. The genus was placed with other species of primitive primates belonging to the Plesiadapiformes (Clemens 1974). This infraorder takes its name from the well-known Paleocene genus Plesiadapis, which is found in both Europe and North America (fig. 10.2). As we will see, there are other scenarios emerging.

There are probably four to six families of plesiadapiformes, one of the more common land-mammal faunas of the Paleocene in North America and Eurasia. Some prefer to call all plesiadapiformes archaic primates while reserving the term euprimates (primates of modern aspect) for

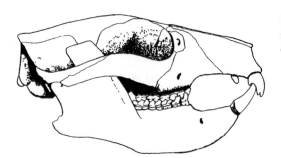

10.2 Skull of *Plesiadapis tricuspidens*. (Redrawn from Gingerich 1976)

those taxa found from the early Eocene and beyond. It is somewhat difficult to morphologically characterize these early Paleocene mammals into discrete taxa because the farther back we go in the fossil record the nearer we get to common ancestors and the more vague become the boundaries among them. Also, the fossil record becomes more fragmentary and often only teeth remain, as in the discovery and identification of the first specimens of *Purgatorius*.

Plesiadapiformes have an auditory bulla formed by the petrosal portion of the temporal bone (a hallmark of living primates) and a dental formula in most species of 2-1-3-3/2-1-3-3, for both upper and lower jaws, indicating a reduction from the basic eutherian 3-1-4-3/3-1-4-3 dental formula (see chap. 5). The teeth of these archaic "primates" had evolved molar features (lower, rounder cusps) suggesting a diet of leaves and fruits rather than strictly of insects. Indeed, two genera had extreme dental specializations, as in *Carpodaptes*, which had evolved elongated, serrated upper and lower premolars, especially the lower P4, which had anywhere from four to nine separate cusps (fig. 10.3). These teeth were probably used to crack nuts, seeds, and hard fruits as are the teeth of some Australian marsupials today.

Whether any of the plesiadapiformes are actually primates or not is currently undergoing intensive study, and new ideas and theories are beginning to emerge regarding the primate legitimacy of these early mammals. As we will see in the following pages, there is no longer consensus among the various students studying these fascinating forms. This is an exciting time for paleoprimatologists, but a difficult time to be attempting a synthesis.

The evidence that is promoting the controversy is coming mainly from the fossils themselves, is both cranial and postcranial, and suggests that plesiadapiformes may be related to the modern colugo, *Cynocephalus*, the flying lemur of southern Asia and the Philippines (Beard 1990, Kay

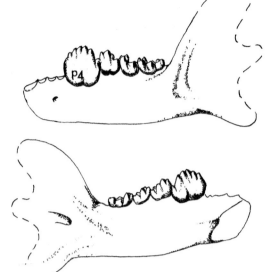

10.3 The lower jaw of *Carpodaptes* from the Late Paleocene of North America, showing the serrated crown of the lower P4. (Redrawn from Szalay 1976)

et al. 1990, and Martin 1990). (Incidentally, flying lemurs are not lemurs, although the face somewhat resembles that of lemurs, but a group of rather specialized mammals which have always presented zoologists with something of a categorizing problem.) The postcranial evidence concerns the anatomy of the hands of several groups of plesiadapiformes, which have elongated middle phalanges that may have served for the attachment of a gliding membrane in a manner similar to its position in the modern colugo (Beard 1990). Beard concluded from this that primates should be combined with Dermoptera (flying lemur) in a new super-taxon, Primatomorpha.

The cranial evidence comes mainly from the ear region in a nearly complete cranium found of *Ignacius*, a plesiadapiform from the Paleocene of Wyoming (Kay et al. 1992). These investigators found that the bone forming the auditory bulla in *Ignacius* was the entotympanic rather than the petrosal from the temporal bone. Because this latter condition is found in all primates and, as mentioned above, is considered a hallmark of primate status, the authors suggested that the plesiadapiformes may be the sister taxon of the flying lemur rather than of primates, and that the order Scandentia (tree-shrews) is the proper sister group of the order Primates.

In his excellent review of the plesiadapiformes, Gunnell (1989) included them in Primates as a suborder. At that time he believed plesiadapiformes represented an evolutionary grade close to that of modern tree-shrews, "not quite insectivore, yet not quite primate either" (p. 147). In his book on primate evolution, Conroy (1990) referred to plesiadapiformes as "one of the most successful of early primate families in terms of both diversity of species and abundance of specimens" (p. 68). At the same time, however, he recognized the difficulty in supporting their primate status and concluded that the position of plesiadapiformes in mammalian evolution will probably remain unsettled for some time. Recently, however, one of the original discoverers of *Purgatorius*, Leigh Van Valen (1994), after detailed comparisons with the dental morphotype of the euprimates, still considers *Purgatorius* a member of the plesiadapiformes, representing the basal radiation of the primates.

In sum, plesiadapiformes were a successful group of Paleocene land mammals and by the late Paleocene made up some 39 percent of the fauna. They included animals ranging in body size from that of living squirrels to that of the largest prosimians today. Their dentitions were morphologically quite variable, suggesting a catholic diet that included insects, leaves, seeds, fruits, and nectars, not unlike the foods eaten today by prosimians. Because of such dental, body-size, and dietary diversity in these Paleocene plesiadapiformes, it is becoming more difficult to maintain that the earliest primates were strictly insectivores who gradually shifted to a fruit and leaf diet (Conroy 1990). Plesiadapiformes were probably nocturnal, although some of the larger forms may have been diurnal. Locomotor habits are not known for all species, because postcranial material is often fragmentary. It appears that some species were probably terrestrial while others (e.g., *Plesiadapis*) were probably quadrupedal and arboreal.

Of the many taxa in the Paleocene, only a few survived into the Eocene, and there is only tentative evidence that any of these were ancestors of the later euprimates. This evidence is dental and relates to the similarity between the teeth of some archaic forms (*Plesiadapis*) and such euprimates as the Eocene *Notharctus* (Jones et al. 1994). Explanations are plentiful regarding their demise in the Eocene. One of the more common theories is that the plesiadapiformes were in competition with rodents and lost out during the early Eocene when the latter groups underwent an explosive radiation. At the same time, environmental and climatic changes in the Eocene undoubtedly played roles, and as with most extinction hypotheses, there were probably multiple causes for the disappearance of these animals. Certainly after the Eocene they were replaced by rodents and primates of modern aspect, whatever the cause of their demise.

I have spent some extra time discussing the plesiadapiformes because they are important to our understanding of the early evolution of several lineages of land mammals closely related to primates, regardless of their exact taxonomic position, which now appears to be nonprimate. The latest consensus is that they are not primates (but see Van Valen 1994), but represent either a sister group to the primates or a "primate-like adaptive group." The arguments and hypotheses will probably continue for years, since debate is a most important and vital part of science. Without such differences of opinion science would tend to stagnate, and certainly it would not be as much fun. The important problem is not so much whether plesiadapiformes are, or are not, primates, but rather what they tell us about the early evolutionary history of archaic mammals. Philip Gingerich, a paleontologist at the University of Michigan, made a rather perspicacious observation regarding these Paleocene mammals when he wrote, "The evolutionary importance of *Plesiadapis* and its allies, like that of the tree-shrews, would be little diminished if they were to be excluded from the Primates, but I think in the case of *Plesiadapis*, as in the case of tree-shrews, we would diminish our understanding of primate evolution by excluding them and the perspective they provide on the form and grade of early primates" (Gingerich 1986, pp. 43–44).

Eocene Epoch, 55–35 mya

There is no doubt that the Eocene was the time of maximum warming during the Cenozoic era. The evidence comes from many sources, including paleoclimatology, paleobotany, and biogeography. Temperatures peaked in the middle Eocene with tropical weather predominating in the northern latitudes, but by the late Eocene, some 35 MYA, temperatures had begun to drop precipitously, and with the lowering temperature came a change in floras, faunas, and geography. Indeed, the transition from the Eocene to the Oligocene is known as the Grande Coupure, or "great cut," since so many mammals became extinct during this period. The result was the virtual disappearance of primates from the northern hemisphere and their appearance in the equatorial regions, "the nadir of primate diversity in northern continents" (Gingerich 1986, p. 33).

During the early Eocene, North America was connected by several islands with Europe and also maintained a close connection with Asia via Beringia. These geographical connections permitted faunal migrations in both directions, and it is estimated that as much as 50 percent of mammalian faunas were similar in both North America and Europe at this time (Conroy 1990). However, by the late Eocene the separation between North America and Europe was complete, leaving the two continents biogeographically (genetically) isolated for the first time in millions of years.

Not only primates proliferated during the first half of the Eocene, but many other groups of land mammals appeared at this time. Indeed, it is estimated that "70% of North American Eocene mammalian genera and 90% of European Eocene genera made their first appearance in the early Eocene" (Conroy 1990, p. 98). It has recently been suggested that the changes in biota (all plants and animals of a given region or time) and temperature occurring at the Paleocene-Eocene boundary were the result of plate-tectonic activities and an increase in carbon dioxide that resulted in global warming (Rea et al. 1990). As continental climates became warmer in higher latitudes, seasonal changes began to appear, and the total result of these new environmental regimes appears to have favored the evolution of most orders of modern mammals (Gingerich 1989).

The Eocene is the setting for another current problem in primate phylogeny: the euprimates, primates of modern aspect, with nails rather than claws, a postorbital bar, and an opposable first toe. The euprimates belong to the families Omomyidae and Adapidae, the two major families making up the primate fauna during the Eocene. The omomyids are generally considered to be tarsierlike; the adapids are more like lemurs in their morphology. We call this a "current problem" because there has been a recent increase in debates and discussions concerning the "old problem" of the phylogenetic relations of the various taxa in these two families (Culotta 1992, Martin 1993, and Fleagle and Kay 1994). The two families are well represented by several genera, and many species are found in North America, Europe, and Asia. Both disappear in Europe by the Oligocene but remain in North America and Asia into the Miocene, albeit rare.

The members of these two primate families are well represented in the fossil record by teeth and cranial and postcranial material. Several genera will be briefly discussed before we synthesize and assess the new information emerging about the origins and radiations of these Eocene primates.

The members of the Adapidae lived in North America, Europe, and Asia and may be divided into two subfamilies, the Notharctinae and the

Adapinae. *Notharctus*, *Pelycodus*, and *Smilodectes* are well known notharctine genera from western North America, while *Cantius*, one of the earliest notharctines, was present in both North America and Europe. Adapines, on the other hand, were limited mostly to Europe, although two genera have been found in Asia and one in North America. At present, *Cantius* appears to be the most primitive notharctine and may represent the ancestral stem from which later notharctines and adapines evolved. Adapines disappear from the fossil record in a somewhat staggered manner beginning in the middle Eocene, and by the late Eocene they had completely disappeared in North America. In Europe, they persisted into the late Eocene, while in Asia, they lasted until the late Miocene before becoming extinct.

The adapids ranged in size from small, probably insectivorous animals weighing no more than about 150 grams (5 ½ oz) to large, certainly folivorous primates weighing some 6 to 8 kilograms (13 to 14 lb). In modern primates a relationship exists between body weight and diet, often referred to as Kay's threshold, which suggests that insectivores typically weigh less than 500 grams while primates weighing more than this are folivores or frugivores (Kay 1984). The body weight of these early primates can be estimated using the correlation that exists between tooth size and body weight (Gingerich et al. 1982 and Conroy 1987). The majority of adapids weighed more than 1 kilogram, especially during their later radiations, which indicates that they were mainly folivorous-frugivorous primates. Their activity patterns were mostly diurnal, except for the European genus *Pronycticebus*, which is believed to have been nocturnal.

Based on most postcranial proportions, adapids were vertical clingers and leapers, especially *Notharctus*, the most lemurlike of all genera. This genus possessed long hindlimbs compared to short forelimbs, and each extremity was equipped with opposable thumbs and big toes that indicate powerful grasping ability.

Other general features of adapids include fused mandibular symphyses, except in *Pelycodus* and *Cantius*. The dental formula is primitive: 2-1-4-3/2-1-4-3 (note the retention of four premolars). The lower incisors are small and vertically implanted, with no suggestion of a tooth-comb as in modern lemurs. The canines are projecting and sexually dimorphic in most species. In fact, Gingerich (1981) uses the degree of canine- and body-size sexual dimorphism in *Adapis* to speculate that these primates lived in polygynous, multi-male groups, which, if correct, extends this mode of behavior back to the Eocene.

It is clear that adapids share many anatomical features with modern strepsirhines, particularly lemurs, but the exact relations have yet to be worked out. Indeed, some students believe adapids may be ancestral to later anthropoids (Gingerich 1980 and Rasmussen 1986).

The Omomyidae are separated into three subfamilies, Anaptomorphinae, Omomyinae, and Microchoerinae, that lived for about 30 million years. There are numerous genera assigned to each subfamily, and like adapids they first appear in the early Eocene of North America, Europe, and Asia. They survived through the Eocene into the Oligocene, and at least one genus, *Ekgomowechashala*, may have been the last member of the omomyids to become extinct in North America in the late Oligocene or early Miocene. The Euroasian omomyids persist into the Oligocene. A fossil lower jaw fragment *Afrotarsius*, discovered in 1985, in the early Oligocene Fayum badlands of Egypt, has been tentatively aligned with the extant Tarsiidae rather than with the omomyids because of the marked similarity between the lower molars of the fossil and those of living tarsiers (Simons and Bown 1985). If the diagnosis holds, this will be the first and oldest tarsiiform primate from Africa (Simons 1995).

The omomyids were tarsierlike, rather small at 10 to 550 grams (1 to 20 oz), insectivorous primates with the large eyes that indicate nocturnal habits in most taxa (Gingerich 1984). Several genera, however were larger, falling above Kay's threshold of 1 kilogram, which suggests a more folivorous diet. Some of these may have been diurnal (Conroy 1990). There is no evidence of sexual dimorphism in tooth or body weight, which suggests that omomyids may have lived in small monogamous groups.

The dental formula is 2-1-3-3/2-1-3-3, except in *Teilhardina*, which retained the primitive four premolars. It should be noted that *Teilhardina* has been found in the earliest Eocene deposits in both Belgium and North America. This may indicate that this early euprimate was very near the base of the radiation that led to all extant primates with the exception of lemurs and lorises.

Omomyids were arboreal primates with long hindlimbs compared with the forelimbs. The distal ends of the tibia and fibula (the two bones of the leg) are fused, and the foot is elongated. These and other postcranial features all indicate a vertical clinging and leaping mode of locomotion, as in modern tarsiers.

Because many of the anatomical features just discussed are similar to those present in *Tarsius*, many scholars classify omomyids with the infraorder Tarsiiformes. However, there is much discussion going on at the present time regarding the taxonomic position of Eocene primates. Notwithstanding, it is understandable why the Eocene is considered the time of maximum prosimian radiations. Many authorities believed until recently (see below) that all later primates emerged from these radiations somewhat as follows: the lemurs and lorises (strepsirhines) arose from the Adapidae, while the tarsiers, monkeys, and anthropoids (haplorhines) evolved from the Omomyidae. Numerous hypotheses in addition to the adapid/omomyid dichotomy have been offered through the years. For example, Fleagle (1988, p. 319) believed that "none of the North American or European prosimians from the Eocene seem very closely related to living strepsirhines, *Tarsius*, or anthropoids." More than twenty years ago, the late British primatologist John Napier (1970, p. 104), wrote that the Eocene radiations represented "the first great parting of the ways which separated higher primates from lower primates." But did they?

In May 1992, a workshop was held at the Duke Primate Center, Durham, North Carolina, in an attempt to come to some decision on this long-standing issue in paleoanthropology (Culotta 1992). The workshop included some of the leading authorities in the field, and as is usual in such gatherings, no single solution was forthcoming after several days of discussion. Perhaps the greatest surprise to emerge from the meeting was the proposal that a mysterious "third primate group" may have given rise to the anthropoids rather than either the omomyids or adapids, and that the anthropoid lineage may extend back as far as 60 to 50 million years into the dim past of the early Eocene, or even into the Paleocene. These early fossil primates come from China and (according to Beard et al. 1994) may represent basal anthropoids (see below). What is equally interesting is that the home of origin of primates may more likely be Africa or Asia, rather than North America or Europe, where the majority of early primate fossils have been found. Also, the Burmese late Eocene taxon *Amphipithecus*, long considered a higher primate, its anthropoid status based on its low-crowned molars and deep mandibles (Colbert 1937), may have to be reconsidered in the light of recent finds of more *Amphipithecus* fossils (Ciochon et al. 1985).

The extended time depth of from 60 to 50 million years for the origin of the anthropoids (if accepted) is one of the more exciting ideas to come out of the workshop. It should be mentioned, however, that there were several devil's advocates at the workshop, and the old hypotheses are not dead yet. For example, Gingerich, after examining the material from China, stated that he thought the fossils in question were "hedgehogs," not anthropoids, while Fred Szalay, of Hunter College, thought that the early anthropoids may indeed have come from Africa or Asia, but they will have looked like omomyids.

Since the Duke meeting, Beard and his colleagues have continued their excavations in southeastern China near the village of Shanghuang (Beard et al. 1996). To date they have uncovered more than 75 fossil primates, mainly teeth and jaw fragments, in addition to thousands of other mammal bones. Unfortunately, these fossils are found in rock fissures and there is no volcanic rock for dating, which means they must rely on comparisons of the mammalian fauna with that from North America, which places them at about 45 MYA (Eocene). Several new fossil primate species have been identified belonging to the adapids and the tarsierlike omomyids, the forerunners of the anthropoids. The new simian species *Eosimias sinensis* displays several morphological characteristics of the jaws and teeth that Beard and his colleagues claim distinguish it from adapids and omomyids. The important paleoanthropological point here is that these investigators argue that *Eosimias* represents a third fossil primate taxon separate from the adapids and omomyids that gave rise to the anthropoid primates. Adapids and omomyids have not been overlooked however; they too have been identified in this unusual and important fossil assemblage. These claims will remain controversial, based as they are on jaw fragments and teeth, and may prove premature. More material is required before a final decision can be made with respect to their taxonomic position. And that has happened: more material has been discovered by Beard and his associates from the middle Eocene of China (Beard et al. 1996), and a new eosimiid species, *Eosimias centennicus*, has been described. This new fossil material consists of the right and left lower jaws with teeth (fig. 10.4). The lower canine is robust and projects above the other teeth, and there are three lower premolars (P2-4). The lower molars possess a well-formed trigonid with a relatively large paraconid well separated from the metaconid. The conclusion reached after studying this new fossil is that the Eosimiidae represent a third group of early Cenozoic primates, indicating that the anthropoid clade was distinct by the middle Eocene, and therefore Beard and his colleagues "conclude that *Eosimias* is indeed a basal anthropoid" (1996, p.84). There seems to be little doubt that they are finding what may prove to be some of the most important fossils for unraveling the early radiations of the primates.

The most recent specimen to add fuel to this debate is a skull found recently in the Oligocene (36 MYA) Fayum Depression of Egypt (Gibbons 1994). The skull is apparently not an anthropoid, because it lacks

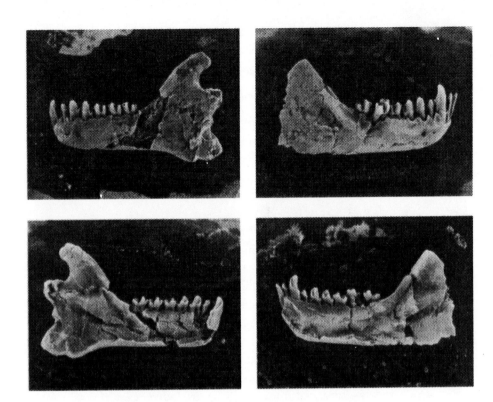

10.4 The lower jaws of *Eosimias centennicus* from China. A) Left side, buccal view; B) Left side, lingual view; C)Right side, buccal view; D) Right side, lingual view. Original about 1 inch long. (Photo courtesy of K. Christopher Beard)

certain anthropoid characteristics: for example, there is no bony septum behind the eye socket and it has four premolars. The skull is so different from other contemporary material that it is given a new genus and species name, *Plesiopithecus teras*, belonging to a new family, Plesiopithecidae, and a new superfamily, Plesiopithecoidea. Although P. *teras* is not an anthropoid, it is an old and very odd prosimian, and as Gibbons notes, "it is bound to make paleoanthropologists take a look at Africa as the possible cradle of the primates" (1994, p. 541). We might also add Asia, as a "possible cradle" of primate evolution, especially in view of *Eosimias*.

Perhaps Robert Martin, of the University of Zurich, has written the most succinct review of all the reappraisals of primate evolution going on at the present time (1993). He concludes that the earlier emergence of anthropoids now seems likely, but that their origins still remain obscure. He further suggests that more caution is necessary when interpreting fragmentary fossils, especially since so often there are only a few specimens to study. He correctly discusses the importance of identifying convergent evolution during the course of primate history as a way of reducing the amount of misinterpretation in the future.

This brief discussion should make it clear that paleoanthropology is alive, complex, and as contentious as ever in the last decade of the twentieth century. There are many more early fossil primate species waiting to be discovered, perhaps by a reader of this book.

Oligocene Epoch, 35 – 24 mya

The climate of the Oligocene continued to become cooler and drier, a trend that had started in the late Eocene as a result of migrating continents that led to India colliding with the Asian mainland and the separation of South America and Australia from Antarctica. These Oligocene events, in turn, caused a general cooling of the ocean waters, which in time began to limit the high-latitude tropical regions that had been common during the Eocene. Eventually the sea level was lowered, due to the formation of ice packs around the poles. It is important to note that Africa was separated from Eurasia by the Tethys Sea during most of the Oligocene, which meant that migrations of land mammals between these two continents were probably impossible, at least until the early Miocene. Also, by the early Cretaceous, Africa was removed from South America with the formation of the South Atlantic sea floor, a process that continues to this day. In addition, North and South America were relatively isolated from each other from about the Cretaceous/Paleocene transition to the formation of the Isthmus of Panama some 6 mya (Emiliani et al. 1972).

The Oligocene was an important period in the history of primate evolution, even though our information regarding this critical time is, with few exceptions, limited to Africa and South America. As noted earlier, the Eocene/Oligocene transition is often referred to as the Grande Coupure (great cut) for a good reason: at least 60 percent of the land mammals of Europe disappeared at this time, as well as most of the North American primates (Savage and Russell 1983).

To date, no fossil prosimians have been found in South America. Many orders of mammals are present in the Paleocene and Eocene formations of South America, but prosimians are not part of this biostratigraphy. The earliest fossil primates are found in the late Oligocene, and all are anthropoids. The first primates appear along with the cavia-morph rodents, a porcupinelike group of mammals that also include guinea pigs (Hoffstetter 1974). Where these animals came from and how they got to South America "are two of the most fascinating and difficult questions in primate evolution" (Fleagle 1988, p. 343). These problems have been hotly debated and have been the bane of paleoan-thropologists for years (see Ciochon and Chiarelli 1980 and Cachel 1981). Unfortunately, there is no indication of a solution in the imme-diate future.

Anthropoid primates appear in the late Oligocene in South America, but did they come from North America, Africa, or perhaps Asia? They had to come from one of these three continents since they were the only sources that had Eocene primates. Of course there is Antarctica, a far warmer place then than it is now, but as far as we know there were no Eocene primates living in Antarctica. Until the Oligocene there is no positive evidence of anthropoid primates anywhere in the world unless we accept the material from China that Beard and his colleagues have proposed as simian primates (1996). This has recently been supported by Kay et al. (1997) who also consider these China fossils to be prim-itive anthropoids. And of course there is the Eocene *Amphipithecus* mate-rial from Burma that has been referred to as anthropoid (Ciochon et al. 1985). There are several scenarios that attempt to explain how anthro-poid primates arrived in South America; each has its advocates, as well as its detractors.

One thesis holds that New World monkeys had their origin in Africa and rafted across the South Atlantic (Hoffstetter 1974 and 1980) some-time during the early to middle Oligocene even though these two con-tinents were 500 to 3,000 kilometers apart by the end of the Eocene (Conroy 1990). Another theory is that the anthropoid grade of primate evolution was achieved in North America and that then these primates migrated to the warmer climate of South America. An interesting twist on the last theory is that of McKenna (1980), who suggested that both primates and cavimorph rodents were conveyed on islands that were themselves moving south between North and South America ("Noah's Arks") due to tectonic forces. At this writing there is no clear answer as to how New World monkeys got to their present home in South America, but evidence is fast accumulating to support the thesis that platyrrhines came from Africa.

Another interesting question about this epoch is biological: is the suborder Anthropoidea monophyletic (arising from one group) or diphyletic (arising from two different groups)? To answer this question it is necessary to determine if platyrrhines and catarrhines had a common ancestor before they separated. To our knowledge, this question has not yet been satisfactorily answered, although the new Eocene primate fossils from China and the Fayum of Egypt may hold answers to some of these most intriguing and fascinating questions in primate paleontology (Simons and Rasmussen 1994). Also, we must not forget *Amphipithecus* and *Pondaungia* from the Eocene of Burma as possible candidates for anthropoid status. Even allowing the anthropoid status of any of this material, there is still the original problem of getting the animals to South America in the Oligocene.

There are three late Oligocene genera from South America: *Branisella* from Bolivia, and *Tremacebus* and *Dolichocebus* from Argentina. The fossils are all fragmentary, consisting mainly of jaw parts and damaged skulls, but all have a platyrrhine morphotype.

In 1983 new specimens of *Tremacebus* and *Dolichocebus* were collected from two sites in Argentina, the first primates to be recovered from these sites in over forty years (Fleagle 1985). The material is all quite fragmentary, but Fleagle was able to conclude that the several dental and postcranial similarities link the Oligocene anthropoid primates of the Fayum (to be discussed below) with New World monkeys and that this in turn provides "very strong morphological evidence that the earliest platyrrhines are derived from an African anthropoid stock, rather than from North American prosimians" (Fleagle 1985, p. 271).

The Fayum Depression is located some 150 kilometers (90 mi) southwest of Cairo. Today this area is a desert badlands that has been a paleontological gold mine since before the turn of the century (fig. 10.5). It is the most fossiliferous region in the world for Oligocene primates, and more recently Eocene primates have been found in the lower formations of this amazing and seemingly unending supply of fossil primates. The most recent statement regarding the Fayum Eocene primates is that "it is now apparent that there was a radiation of small-bodied, fruit-and-insect eating anthropoideans during the Eocene" (Simons and Rasmussen 1994). There are several good accounts of the long history of paleontological expeditions to the Fayum (e.g., Simons and Wood 1968 and Conroy 1976). Some of these were before the turn of the century, but it was not until the early part of this century, when the American Museum of Natural History began its excavations, that fossil primates started to be discovered. In 1961 Elwyn Simons, then at Yale University, now at Duke University, began systematic excavations in the

10.5 A view of the Fayum, Egypt. The Fayum has been rich in Oligocene fossil primates since its discovery in the early part of the twentieth century. (Author's photo)

Fayum that are still going on today. He and his dedicated fossil hunters (students, colleagues, and an occasional pilgrim) have discovered a wealth of specimens that have increased our knowledge of anthropoid evolution during this critical period. Simons (1995) has written an excellent review of the Fayum primates that covers the paleontological work carried out there from 1908 to the present.

Thirty to thirty-five million years ago the Fayum was quite different. During the late Eocene and Oligocene it was a tropical, moist region, mostly swampy, with alternating wet and dry periods (Bown et al. 1982). Trees were common around the borders of the swamps and streams, and the climate supported a large and diverse fauna and flora. There are many fossil trees on the surface of the desert floor today that are autochthonous, attesting to the forested nature of the region during the Oligocene. The sediments that were being deposited during this time make up the various layers known as the Jebel Qatrani Formation, the source of the Fayum fossils.

The Parapithecidae are the most common primates from the Fayum. The family contains the smallest anthropoid primate yet discovered, known as *Qatrania*. It weighed only about 300 grams (less than a pound) and possessed a molar morphology suggesting a frugivorous diet, in spite of its small size. The other two genera of the family are larger than *Qatrania*, but parapithecids are still smaller than the propliopithecids. Most parapithecids have a primitive dental formula 2-1-3-3/2-1-3-3 and have retained the third premolars, a feature common to all modern platyrrhines. Interestingly, *Parapithecus grangeri* has lost its lower permanent incisors, a unique condition in primates since it is the only known primate lacking both lower incisors.

Parapithecids possessed a round ectotympanic bone attached to the bulla, as in cebids, and not the tubular configuration of all Old World anthropoids. In a detailed study of the auditory region in primates, Conroy (1980) found this region in the Fayum primates to be more similar to that of New World monkeys than to that of any of the other groups studied.

The anatomy of the limbs indicates that parapithecids were arboreal quadrupeds, but with good leaping abilities. In many ways the postcranial anatomy of parapithecids is more like that of New World monkeys than it is like that of Old World monkeys or apes.

The Propliopithecidae are larger than the parapithecids, with *Aegyptopithecus* being the largest Fayum primate, 6 to 8 kilograms (13 to 17 lbs) (Fleagle 1988). The fossil remains of *Aegyptopithecus* and *Propliopithecus* have been studied by many students through the years and are pretty well known (fig. 10.6). These early anthropoids are often referred to as apelike primates, and certainly they show a mosaic of both ape and monkey characteristics. Their earlier allocation with living apes is understandable since it was based primarily on dental characters, and the lower dentition of *Propliopithecus* and *Aegyptopithecus* certainly presage the dentition of later apes. The mandibular symphysis is fused in both of these Fayum families.

With the continued acquisition of cranial and postcranial material, most of these Fayum anthropoids seem more platyrrhinelike. There are exhaustive lists and analyses of the many different and similar character states held in common by these species (see especially, Fleagle and Kay 1987, Fleagle 1988, and Conroy 1990).

Most scientists currently agree with the assessment that the Fayum anthropoid primates phylogenetically represent a group of small-to-medium-sized animals who ate more fruits, seeds, and gums than they did leaves. They were arboreal quadrupeds that could leap when necessary and probably rarely came to the ground. Since the parapithecids had sexually dimorphic canines it is likely that they were polygynous. All Fayum primates were more primitive than later anthropoid primates.

10.6 *Aegyptopithecus zeuxis*, showing the cranial base region as well as the beautifully preserved teeth in the maxilla. Dental formula: 2I-1C-2P-3M. (Cast from Carolina Biological Supply Co.)

The parapithecids appear to be the most primitive of the Fayum primates and therefore more like the smaller living platyrrhines of South America. Thus, they predate the platyrrhine-catarrhine split. The propliopithecids, because they are more advanced than platyrrhines but more primitive than later catarrhines, are the likely candidates for the ancestors of the catarrhines, that is, they represent primitive catarrhines before the split into apes and Old World monkeys. If this phylogenetic assessment proves to be correct, it appears that Africa is the home of origin of the higher primates.

MIOCENE EPOCH, 24–5 MYA

In general, the climate of the Miocene was somewhat warmer than that of the preceding Oligocene, but toward the end of the Miocene it was cooler and drier in the low and middle latitudes of the Old World. Africa and Europe converged and made contact with each other while

India continued to collide with Asia, forming the Tibetan Plateau and the Himalayan mountains. Both of these tectonic events helped to reduce the Tethys Sea. Incidentally, a few million years after Africa pushed up highlands across the narrow Strait of Gibraltar and closed off the western mouth of the Tethys Sea, the Mediterranean went totally dry. This occurred about six million years ago and formed, for approximately a half million years, one of the more inhospitable environments ever known (Raymo 1983). This barren, hot desert region may have been a more formidable barrier to our primate ancestors than the former sea.

Whatever the final decision is concerning the Oligocene "apes," it seems most likely that the Miocene apes had their origins from one, or more, of these Oligocene lineages. The Miocene apes spread rather quickly over much of the Old World and became the dominant primate for a brief period of time. Before the end of the Miocene the apes were extinct in many of the areas they had previously inhabited. This extensive Miocene radiation was responsible for the apes of today even though it is not always clear which lineages were the ancestors of which apes. For example, there are no known fossil ancestors of the chimpanzee or gorilla, and not all authorities agree on the fossil history of the orangutan.

A recent article appearing in *AnthroQuest* (1992) was titled "The Muddle in the Miocene," which accurately reflects the present understanding regarding the biological diversity of Miocene hominoids. As this paper points out, finding more (new) fossils does not necessarily mean a clearer understanding is forthcoming. Rather, the plethora of new specimens uncovered over the past decade has only revealed the great adaptive diversity of Miocene apes. There now appear to be literally dozens of apelike primates in the Miocene who had spread over much of equatorial and subequatorial Africa and Eurasia, many with different specialized adaptations, habits, and life styles.

Glenn Conroy (1990) offered a classification of this diversity into three groups: dryomorphs, the early and middle Miocene hominoids of East Africa and Eurasia with thin enamel on their molar teeth; ramamorphs, middle Miocene hominoids of East Africa and Eurasia with thick molar enamel; and pliomorphs, early and middle Miocene hominoids of Eurasia possessing many shared primitive catarrhine characteristics. Table 10.2 presents Conroy's classification of Miocene hominoids. It should be mentioned that several authorities, including Conroy, place *Ramapithecus* and *Kenyapithecus* within the genus *Sivapithecus*.

Table 10.2   Classification of Miocene Hominoids

Taxa	Epoch and Region
Superfamily Hominoidea	
Family Proconsulidae	E. Miocene [Africa, Asia]
Dendropithecus	E. Miocene [Africa]
Dionysopithecus	? E. Miocene [Asia]
Limnopithecus	E. Miocene [Africa]
Micropithecus	E. Miocene [Africa]
Proconsul	E. Miocene [Africa]
Rangwapithecus	E. Miocene [Africa]
Family Oreopithecidae	E.-L. Miocene [Africa, Eur.]
Nyanzapithecus	E.-L. Miocene [Africa]
Oreopithecus	L. Miocene [Eur.]
Family Pongidae	M. Miocene-Pleist. [Africa, Eur., Asia]
Dryopithecus	M.-L. Miocene [Eur.]
Gigantopithecus	L. Mio.-Pleist. [Asia]
Lufengpithecus	L. Miocene [Asia]
Sivapithecus	M.-L. Miocene [Africa, Eur., Asia]
Otavipithecus	M. Miocene [S. Africa]
Family Pliopithecidae	M.-L. Miocene [Eur., Asia]
Laccopithecus	L. Miocene [Asia]
Pliopithecus	M.-L. Miocene [Eur.]

Adapted from Conroy 1990

The great East African Rift System formed during the Miocene and extends today from the Red Sea in the north to Mozambique in the south, a distance of some 3,000 kilometers (about 1,860 mi). Most of the early eastern African Miocene hominoid sites are located in these rifts or valleys, date from 20 to 17 MYA, and include such genera as Proconsul, Limnopithecus, Rangwapithecus, and Afropithecus. The environment of East Africa underwent several major changes, from more forested areas in the early Miocene to more open woodland regions in the middle Miocene (Andrews and Van Couvering 1980 and Pickford 1983). Little is known about the late Miocene climate in East Africa except that it continued to get cooler and drier (Conroy 1990). Thus, early East African hominoids evolved in a forest environment that gradually shifted to a more open-country habitat.

These early Miocene forms varied in size from that of a small monkey to that of a chimpanzee. *Proconsul* lived in tropical forests and mixed environments, and its skeleton indicates that it employed quadrupedal walking and climbing while in the trees, with little suspension from branches. *Proconsul* was sexually dimorphic, and its teeth suggest a diet of leaves and fruits. The middle Miocene saw the evolution of *Kenyapithecus* at several northern Kenya sites, such as Fort Ternan. It was more apelike than *Proconsul* in several dentofacial features and had thick enamel on the molars, which might indicate a shift in diet to a tougher type of food. Certainly they were equipped for robust, powerful chewing and displayed a marked sexual dimorphism in canine size. The recently discovered lower jaw of *Otavipithecus* (fig. 10.7), from the late middle Miocene of Namibia, makes it the first Miocene ape from southern Africa, extends the southern boundary of Miocene hominoids into South Africa, and indicates for the first time that Miocene hominoids occupied the full length of the African continent.

Middle Miocene apes (15–14 MYA) have been found in Eurasia, are mainly represented by teeth from such sites as Pasalar in Turkey and Neudorf in Europe, and appear to be close to *Kenyapithecus* and *Proconsul* in their bevavioral habits. *Dryopithecus* appears somewhat later in western and central Europe (9–11 MYA), and *Ouranopithecus* in late Miocene layers (9–10 MYA) of northern Greece. Their jaws and teeth suggest a diet similar to that of *Proconsul*. They were probably mainly arboreal, spending some time on the ground. A recent discovery in Spain of a partial skeleton of *Dryopithecus laietanus* seems to provide evidence that bipedal posture and locomotion had appeared by 9.5 million years ago (Moya-Sola and Kohler 1996). If this proves correct, it will be the oldest evidence of bipedalism by some 5 million years. The origin of the dryopithecines, however, still remains unknown.

The greatest diversity in genera, as well as in geographical distribution, of Miocene hominoids occurred in the late Miocene. They were spread from Spain (*Dryopithecus*) east through Europe (*Oreopithecus* and *Ouranopithecus*) to Pakistan (*Sivapithecus*) and across the Tibetan Plateau to China (*Lufengpithecus*). And of course there were still various species left in Africa, for example, *Kenyapithecus*.

Sivapithecines are widely distributed in eastern Africa and Eurasia, but the most prolific sites have been in the Siwalik Hills of Pakistan (fig. 10.8). Fossil hominoids were found in these hills before the turn of the twentieth century and are being discovered there today. *Sivapithecus* is perhaps the best-known species, and its fossils have been studied

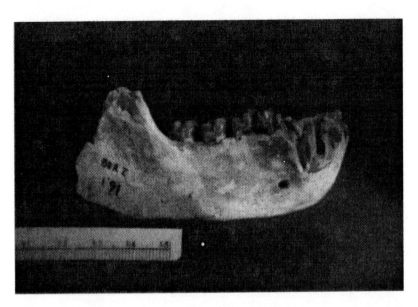

10.7 The lower jaw of *Otavipithecus namibiensis*. (Photo courtesy of Glen C. Conroy)

by many scholars through the years. There is little doubt among most paleoanthropologists that the *Sivapithecus-Pongo* clade is a valid representation of the biological relationship between these two primate taxa. *Sivapithecus* and *Pongo* share several features of the skull and face, such as absence of a frontal air sinus, a narrow bony partition between the orbits, postorbital constriction, and a similar subnasal region (fig. 10.9). Of all Miocene hominoids, *Sivapithecus* may be the only species with living descendants today, the orangutans of Southeast Asia. The genus *Pongo*, however, does not appear in the fossil record until the Pleistocene.

One of the largest primates that ever lived, *Gigantopithecus* (based on dental dimensions) weighed between 150 kilograms and 230 kilograms (230 to 506 lbs) (Conroy 1987). There were two species: the smaller *G. giganteus* was from the latest Miocene of India and Pakistan, while the larger species, *G. blacki*, was living in China and Vietnam during the Pleistocene. They are known only from lower jaw fragments and isolated teeth (thick enameled), which suggest that they ate tough, fibrous plants, possibly bamboo, as do modern pandas (fig. 10.10). Their size would seem to have precluded anything but terrestrial locomotion.

One of the most difficult late Miocene hominoids to classify was found in lignite deposits in Italy dated between 7 and 9 MYA. This is *Oreopithecus*. It was discovered toward the end of the nineteenth century

10.8 The Siwalik Hills of Pakistan, showing the exposures and geology of this region. (Author's photo)

10.9 Facial view of *Sivapithecus indicus* (GSP 15000). Notice the narrow lateral incisors compared with the central incisors. (Cast courtesy of David R. Pilbeam)

10.10 Lower jaw of *Gigantopithecus blacki* from China. (Cast courtesy of Milford H. Wolpoff)

and is now known from many specimens, including almost complete skeletons excavated in the 1950s. Still, there is much debate concerning its taxonomic position: that is, is it a cercopithecoid monkey, or is it a hominoid? A recent review of its status by Harrison (1986) considers it a hominoid due to its shared, derived postcranial features and concludes that its dental similarities with cercopithecoids are the result of functional convergence. *Oreopithecus* is an excellent example of mosaic evolution, which was discussed earlier in this chapter. The genus, however, is still placed in its own family, Oreopithecidae, by some paleoanthropologists (Conroy 1990; see table 10.2).

An important late Miocene site is in the Lufeng Basin in Yunnan Province, China (fig 10.11). During the late Miocene, this region was a moist, but not wet, upland tropical forest with many open glades, according to a detailed analysis of the flora and fauna (Badgely et al. 1988). Certainly, the rising Himalayas were affecting the climate and geography of a wide region around this massive uplift. Moisture-bearing winds, on both the east and west flanks, were controlling the respective watersheds, and at the same time changing geological conditions influenced the routes taken by migrating animals.

10.11 Lignite exposures at the Lufeng site in Yunnan Province, China. Many primate fossils have come from this location during the last three decades. (Author's photo)

Originally discovered in the 1950s, the Lufeng site has proved to be an exceptional late Miocene location, yielding many teeth, jaw fragments, and more importantly, skulls and postcranial material. Chinese paleoanthropologists originally identified two genera, *Ramapithecus* and *Sivapithecus*, but with the addition of so much material through the years and the continued study of these specimens, it is generally agreed that there is only a single highly sexually dimorphic taxon, *Lufengpithecus*, represented at Lufeng (Kelley and Etler 1989 and Conroy 1990; see fig. 10.12). Due to the generally primitive craniodental features in the Lufeng specimens, it is unlikely that *Lufengpithecus* belongs in the *Sivapithecus-Pongo* clade as originally proposed.

The late Miocene hominoids were similar in many ways: they lived in forests or wooded areas; many appear to have been quadrupedal climbers and walkers, but some forms may also have employed suspensory activities; they were moderately large to quite large (large monkeys to small gorillas); and they were all sexually dimorphic. Their diets were undoubtedly quite varied, consisting of rather hard, tough foods (eaten by the animals with the thicker molar enamel) or of softer items such as leaves and fruits (preferred by animals with thinner enamel). It is interesting that, to date, never has more than one hominoid species been

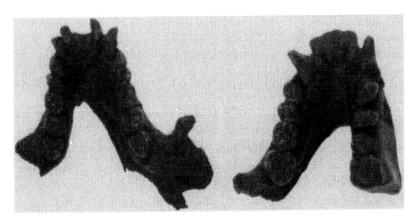

10.12 The lower jaw of *Lufengpithecus*. Female on the left and male on the right. (Author's photo of casts)

found living at the same time at any of these Miocene sites. By the end of the Miocene the majority of these lineages were extinct, with the possible exception of *Sivapithecus* mentioned above. Thus, to date, there are no fossil remains of chimpanzees or gorillas from Africa or Eurasia. The most widely accepted scenario explaining these events is as follows: there is good evidence for a cooler, drier climate followed by the general reduction of thick forests and replacement with more open woodlands; as the Miocene hominoids became extinct, there was gradual replacement by monkeys that were better adapted for life in the more open habitats.

The New World monkeys continued to evolve throughout the Miocene in South America and probably on several islands in the Caribbean Sea as well. Much information regarding platyrrhine evolution comes from the early to middle Miocene of Colombia, and more recently fossils from the early Miocene of Argentina have been uncovered that suggest an adaptively diverse fauna (Fleagle et al. 1987 and Anapol and Fleagle 1988). The callitrichids are represented by three fossils from the middle Miocene of Colombia, whereas cebids are somewhat better known during the middle Miocene and have been found in Colombia as well as in Argentina. An interesting aspect of the fossil record of the platyrrhines is the longevity of several of the lineages. For example, the squirrel monkey, *Saimiri*, and the night monkey, *Aotus*, have histories going back to the middle Miocene (Rosenberger 1994). Rosenberger also notes that the genera *Alouatta*, *Callimico*, and *Pithecia* are only slightly different from several fossils from the La Venta middle Miocene site in Colombia. He believes that these long lineages may reflect "a single adaptive radiation that has continued without much replacement since its origins" (p. 216). As explorations continue in Central and South America we can expect a clearer picture to emerge regarding the origins and adaptive radiations of platyrrhine monkeys.

The history of the Old World monkeys is better known during the Miocene than is that of the New World monkeys. Old World monkeys share many similar characteristics inherited from a common ancestor and appear to be a monophyletic group. Cercopithecoids do not appear until the early middle Miocene, 17 to 15 million years ago, when *Prohylobates* is found in Egypt and Libya and *Victoriapithecus* is represented in Kenya. There is little doubt that these species were cercopithecids, but opinions differ with respect to the time of formation of the two subfamilies Colobinae and Cercopithecinae. The split probably occurred sometime during the middle Miocene. After their appearance the cercopithecoids underwent an expansion into the more open lands that were beginning to appear, and at the same time they probably became more dependent on leaf-eating. Certainly by the late Miocene, cercopithecoids appear to have been taking advantage of the opening habitats by spreading over much of the Old World and replacing many hominoid species. Before the end of the Miocene, Old World monkeys were represented by such diverse forms as *Macaca, Parapapio, Mesopithecus,* and *Libypithecus*. The former two species came from the late Miocene of northern Africa and southern Africa respectively and belonged to the cercopithecines, while the latter two species were colobines from Eurasia and Egypt.

PLIOCENE EPOCH, 5 – 1.8 MYA

The Pliocene terminated with the beginning of the Pleistocene epoch, a time of worldwide glaciation that lasted until about 12,000 years ago. These two epochs are often called the Plio-Pleistocene and considered together. The Plio-Pleistocene ended as the last epicontinental glaciers receded north leaving an essentially modern landscape with modern fauna and flora.

The evolution of the living lesser apes and great apes remains a mystery. As we have seen, gibbonlike fossils have been found in the Oligocene (e.g., the small *Propliopithecus*) and in the Miocene (*Pliopithecus* and *Dendropithecus*). None of these fossils, however, is anatomically a gibbon.

Until recently the fossil record of gibbons began only in the middle Pleistocene of China and Indonesia. A late Miocene form, *Laccopithecus robustus*, from Lufeng, China, is now acknowledged as a true gibbon (Pan Yuerong 1988). The divergence of gibbons from the hominoid line has been variously estimated between 17 and 20 MYA or to 12 MYA by DNA and immunological studies. The Lufeng site has been dated at 8 to 7 MYA, which suggests that the gibbons separated from the hominoid line somewhat later than suggested by the DNA studies. Or perhaps the earlier part of the gibbon lineage remains unknown.

The Asian orangutan is considered by most primatologists as being the descendant of the late Miocene genus *Sivapithecus* of Pakistan (Kelley and Pilbeam 1986 and Kelley 1994). There are also fossil orangutan teeth from the karst caves of south China and Java dating to the Pleistocene (Hooijer 1948 and Ho et al. 1995).

Fossils of the African great apes are unknown. The living chimpanzees and gorillas are not related to any of the known lineages of Miocene apes, but it should only be a matter of time before an ancestor is found in a late African Miocene or Pliocene site. The molecular evidence suggests between 6 and 10 MYA for the separation of the African great apes from the hominoid stem. This puts the split in the late Miocene, which fits pretty well with what many paleoanthropologists believe today.

The Old World monkeys continued to radiate through the Plio-Pleistocene into many new econiches, particularly in Africa and Asia. These migrations contained many new species of both subfamilies of Old World monkeys, but the colobines were the most productive, with as many as seven new species appearing in the late Pliocene (Delson 1994). Some of these (e.g., *Rhinocolobus* and *Paracolobus*) were larger than living colobines, although smaller colobines were also present. Baboons were represented in the African Plio-Pleistocene by *Papio* and *Theropithecus*. Colobines also faired well in Europe and Asia during the Plio-Pleistocene. *Mesopithecus*, a moderate-sized colobine, lived from England in the west, through Europe, to India in the east through much of the Pliocene. During the Pliocene, macaques lived in North Africa and Europe and, sometime in the early Pliocene, expanded westward as far as England and eastward to China and eventually to Japan. There is evidence of macaques surviving in Europe until the middle Pleistocene, which is about the time they arrived in Japan (Iwamoto 1975).

There are two genera that appeared during the Pliocene that are of great interest to us. These are *Australopithecus* and *Homo*, which represent the first members of the family Hominidae. The australopithecines seem to have had their origins in eastern Africa 5 or 6 MYA, or perhaps even earlier, and evolved into several species during the Pliocene. Their immediate antecedents are unknown at the present time but will undoubtedly be found in Africa in one of the late Miocene hominoid radiations. The genus Homo appeared somewhat later, between 2.5 and 2 MYA, in both eastern and southern Africa. There is good evidence that these animals were coeval for more than a million years. The evolutionary trends that appeared in these early hominids have continued to the present day, resulting in our own species, Homo sapiens.

C. Patuellus Horned Monkey

11 ～

Primate

Conservation

Conservation is a subject much discussed today by scientific organizations, civic and philanthropic groups, countries, and governments throughout the world. At the request of the Global Environmental Facility for the United Nations Environment Program, some 1,500 scientists from around the world met in Jakarta, Indonesia, in 1995 to discuss the plight of the earth's fauna and flora. Their conclusion: the animals and plants of the world are being destroyed at an "alarming rate." This august group suggested the following reasons for the decline in species:

1. Increased population and economic development, which deplete biological resources.
2. Human failure to consider the consequences of actions that destroy habitat, exploit natural resources, and introduce nonnative species.

3. Failure of economic markets to recognize the value of maintaining the diversity of species.

4. Increased human migration, travel, and international trade.

5. The spread of water and air pollution.

Homo sapiens is the common factor in each of these statements. Our species today has about 5.7 billion individuals, with an estimated 7.8 billion expected by 2050 (Cohen 1995). We occupy all continents, most environments, and the majority of the earth's latitudes. There is little doubt that we are the dominant species on earth today and that we possess the power to save or destroy the planet. As the well-known naturalist Jim Fowler recently wrote concerning the extinction of species, "We can no longer ignore their true value to Earth's ecosystem, of which we are also a part" (Fowler 1995, p. 8).

Nonhuman primates comprise only a small fraction of the earth's total species population, with approximately two hundred different species today, but fortunately there is a growing concern for their conservation in many countries. This is particularly true in those countries with tropical forests, where about 90 percent of the world's wild primate populations live, because it is those areas that are undergoing the greatest destruction of primate habitats due to timber extraction and clearing forests for human occupation. As the forests disappear, so do the animals that live in them. It is ironic, as primatologist Nina Jablonski has so cogently pointed out, that "primates (and apes, in particular) are threatened because of the very characteristics that were originally the keys to their success" (1995, p. 115; see chap. 2).

According to Mittermeier and Cheney (1987) there are three major threats to wild primate species: habitat destruction, hunting for food and sport, and live capture for export or local trade. Of course these vary from country to country, but one or more of them are operating wherever primates are having problems. Each of these conditions exacts its own toll on wild primate populations, but without doubt the most catastrophic is habitat destruction. Forests are being removed for agriculture, lumber, industrial purposes, and fuel for cooking. Here again, the major impetus underlying the destruction of tropical forests is human population growth. Population growth in most tropical countries is much greater than in other countries, averaging over 3 percent per year. Brazil averages about 2.3 percent, Nigeria 3.3 percent, and Kenya over 4 percent, compared with 0.7 percent and -0.2 percent for the United States and Germany (Mittermeier and Cheney 1987).

Deforestation, for whatever reasons, causes the death of thousands of primates each year: "The survival of 58% of the species of living primates is being jeopardized, primarily by deforestation," wrote Wolfheim (1983, p. 751). At the current rate of destruction, it is estimated that by the end of the twentieth century most of the tropical forests will have disappeared, to be replaced by subsistence farming, permanent farming, and large plantations. Land taken for these purposes is often overworked by continuous use, which leads to erosion and soil depletion. Much of the land becomes poor in production value as more and more is asked of it. The absence of the once lush ecosystem of the rainforest diminishes Earth's ability to process carbon dioxide out of the air and changes the weather cycles for the whole world. The result is not only the loss of many primate species, and a real threat to ourselves, but the loss of thousands of other animals and plant species, many of which have been shown to have potential for us in health and medicine, not to mention the enrichment of our world by variety, diversity, and beauty (Wilson 1992).

One of the largest tropical islands in the world, Madagascar, houses the world's total population of wild lemurs, about half of the world's chameleons, and some of the world's largest earthworms. Some 2,000 years ago, however, humans came across the Mozambique Channel from Africa, and things began to happen. Giant birds, giant tortoises, and a lemur as large as a modern gorilla became extinct by the seventeenth century (Culotta 1995). But were humans solely responsible for these megafauna extinctions? At a meeting held in 1995 at the Field Museum in Chicago, a group of scientists reviewed and discussed the newest evidence from many disciplines regarding these extinctions. According to Culotta (1995), the consensus was that both climatic shifts and agriculturists may have caused habitat loss, which, when added to nonsustainable rates of human hunting activities, probably were responsible for the extinctions. There is little doubt that many lemur species, as well as other animal groups, became extinct following human occupation of the island. Unfortunately, there is evidence that the process continues today. Lemurs are eaten in certain regions of the island, and the aye-aye, one of the world's most endangered primates, is not only hunted and eaten in parts of Madagascar, but is also killed because it is thought to cause bad luck to the natives if it enters their villages. The distribution of the aye-aye is poorly known today, but an encouraging discovery has been made of a group of aye-ayes on the coast of northwestern Madagascar where they were previously unknown (Simons 1993). The Duke Primate Center proudly reported the first birth of an aye-aye outside of Madagascar on April 5, 1992 (Simons 1993). It was a male weighing 136 grams (about 5 oz) when found, and it was doing quite well when the article was written.

In 1995 another important conservation meeting was held at the University of Madagascar to determine which areas of the island should be considered highest priority for conservation efforts. The results of this meeting left no doubts "that Madagascar's conservation policy has a new intellectual foundation" (Jolly 1995, p. 1). It is hoped that other countries with native primate populations will consider such conservation meetings in the future, for it is only from such scientific conferences that important decisions emerge regarding whether endangered fauna and flora will survive to continue life on earth or perish forever. It is largely in human hands. Education is certainly imperative, and in the long run "conservation depends on in-country support and the development of sound governmental policy" (Yeager et al. 1995).

Hunting remains a problem in many countries where wild primates are considered an important food source, for example, in the Amazon, in parts of central and western Africa, in Southeast Asia, and to some extent, in parts of Madagascar. In western Africa there is a large trade in "bushmeat" (monkey and ape meat), which is very popular and is sold in the larger cities. It is estimated that thousands of monkeys in Sierra Leone are killed yearly to supply meat to the Liberian market (Kavanagh 1984). There is evidence that logging increases the hunting of primates because it opens up roads into the forest for the hunters. Today, with the availability of modern firearms, primates in many countries are having a difficult time maintaining their breeding populations. Hunting for skins and other body parts, particularly skulls, for ornamentation such as rugs and coats is a profitable business in many parts of the world. More than thirty years ago the naturalist George Schaller reported that humans kill gorillas for "sport," and unfortunately we are still the chief predator of the species. Killing the mountain gorillas for their skulls and hands in Rwanda and Zaire to sell to Europeans was reported by the late Dian Fossey as recently as 1983. Two more recent incidents occurred in 1994 and 1995. In 1994 the war in Rwanda claimed its first gorilla casualty when a silverback male stepped on a land mine near Lake Ngezi (Steklis 1995). Then in 1995 four mountain gorillas were killed by poachers in Uganda, probably in order to capture an infant (Nsanjama 1995). Unfortunately, it generally happens that during the capture of live primates several animals are killed for every one that survives. It is estimated that there are approximately 600 mountain gorillas surviving in the wild at the present time in Rwanda, Uganda, and Zaire. The western lowland gorilla (*Gorilla gorilla gorilla*) is faring much better than the mountain gorilla (Chivers 1987). There were about 40,000 of them living mostly in Gabon in 1987. Chivers notes that the other subspecies of the western lowland gorilla, *Gorilla gorilla graueri*, lives in isolated pockets in eastern Zaire, with probably between 2,500 to 4,500 individuals.

Capturing wild primates for pets and biomedical research has been a considerable drain on certain wild primate species. Fortunately, the popularity of primates as pets has apparently waned since the 1970s in many countries, and international trade regulations have become more stringent. Primates are difficult to raise, tame, and breed. They also may carry diseases such as hepatitis, tuberculosis, and several virulent types of viruses (e.g., the Ebola virus). What may be a cuddly little monkey frequently becomes a large, dangerous adult. It should also be added that primates do not make good pets for another reason. As we have seen, primates are social animals who prefer members of their own species to humans. Additionally, they all have sharp teeth, which they use with only slight provocation. Fortunately the pet trade in primates seems to be curtailed in the United States at the present time, but it continues to flourish in many other countries. There are presently ongoing task forces made up of researchers, veterinarians, and other concerned individuals to reduce the incidence of primates kept as pets.

Today, the majority of primates used in biomedical research in the United States are raised in breeding colonies. Consequently there is less demand for importing primates from the wild. The use of captive-bred rather than imported primates benefits both laboratory research and natural populations of primates. The National Institutes of Health has chartered and regulated several Primate Centers within the United States that contribute to a wide range of studies and research, including breeding primates for research. Unfortunately, this is not true for all countries that still import primates for research, although international rules have restricted the trafficking in primates over the last couple of decades. India, for example, has not exported rhesus monkeys for years, and with the drafting of the Convention of Trade in Endangered Species of Wild Flora and Fauna in 1973, England and several European countries have agreed to ban commercial trade in endangered species.

In recent years many of the large zoos in the world have sponsored symposia and conferences on conservation which have resulted in many important publications. Also, zoos, scientific societies, and many private and federal funding organizations have provided funds for naturalistic field studies of primates that have contributed immensely to our knowledge of the myriad problems involved in primate conservation. The American Society of Primatologists, the premier organization in the United States for primate scientists, supports several conservation programs with donated funds. In addition, the Wildlife Preservation Trust International, the Fauna and Flora Preservation Society, the National Geographic Society, the African Wildlife Foundation, and the L. S. B. Leakey Foundation support conservation.

One would be remiss not to mention the pioneering studies of the three great apes in their natural habitats by three women scientists. Jane Goodall began her research on the chimpanzee in the early 1960s and continues today, nearly four decades later. Dian Fossey and Birute Galdikas started their investigations of the mountain gorilla and orangutan in the early 1970s. Dr. Galdikas continues to visit her site, Tanjun Putny Reserve in Indonesia, on a regular basis (Crockett, personal communication). Unfortunately, Dr. Fossey was murdered in her cabin at the Karisoke Research Center in Rwanda in 1985. All three have expressed deep concern for primate conservation, and their publications and public lectures through the years have brought a better understanding and appreciation of the great apes and their endangered plight to the general public. Some believe that the mountain gorilla would not have survived to the present day without the tireless efforts of Dian Fossey to save these magnificent creatures.

Future conservation strategies must consider the local economic situation. Forest conservation is not a high economic priority for most developing countries, yet deforestation, as discussed above, is one of the major reasons for the loss of primate species. Creating large national parks and wildlife reserves is fine, but how many countries can sustain these economically? Myers (1979), in a most interesting book, points out that Tanzania has set aside about 9 percent of its land for wildlife sanctuaries, which is equivalent to the states of Washington, Oregon, and California. People must be made more aware of their relationship with the natural world and how to coexist with it in multiuse areas. Such areas can be used, but not used up. It is also important that field workers be aware of the sensitivities and cultural mores of the local people and not appear as neocolonialists or missionaries spreading their own brand of "truth" (Yeager et al. 1995).

The principle reference work on endangered species is the *Red Data Book*, produced by the Conservation Monitoring Centre of the International Union for Conservation of Nature and Natural Resources in Cambridge, England. In addition, excellent discussions of the serious emergency regarding primate survival in the wilderness can be found in the following books listed here by author and year but appearing as complete citations in the bibliography:

Arambulo et al. 1993
Itoigawa et al. 1992
Kavanagh 1984
Mittermeier et al. 1992
Norton et al. 1995
Smuts et al. 1987
Wolfheim 1983

There are of course many articles written in primatology, anthropology, and zoology journals about the plights of specific primate species that are available in our university and public libraries.

There are some 182 species of nonhuman primates living today. Of these, about 128 species are monkeys, of which 77 are Old World monkeys and 51 are New World primates. Prosimians comprise about 41 species, while apes are usually grouped into 13 species (Kavanagh 1984). In 1983 it was estimated that about one-third of these species were on the endangered species list (Wolfheim 1983).

It is quite apparent that nonhuman primates depend on the tropical forests for their very lives. Most nonhuman primate species are arboreal creatures living in tropical rain forests and are dependent upon the forest and its products for their various life styles. Some species are terrestrial, in that they forage on the ground during the day, but even they return to the trees at night for protection. According to Richard (1985), the majority of nonhuman primates spend from 40 percent to 80 percent of their feeding time eating fruit, leaves, seeds, herbs, insects, gums, and saps. It is both clear and disturbing that destruction of these habitats is eliminating thousands of primate species each year. Unfortunately, one other primate is responsible for most of this deforestation and that is *Homo sapiens*.

⌣ Glossary

ADOLESCENT GROWTH SPURT A period of accelerated growth that occurs in both sexes of some primates, particularly in humans.

AGE-GRADED GROUP A primate social group composed of several adult males and females who differ in status according to age.

ALIMENTARY Pertaining to food, digestion, or the digestive tract (gastro-intestinal tract, or G-I tract).

ALLOMETRY The relative growth relationships between different parts of an organism whereby some parts are disproportionately related to other parts; the study of such relationships.

ALLOPATRIC SPECIES Species living in nonoverlapping geographical regions. Compare SYMPATRIC SPECIES.

ALTRICIAL STRATEGY Species giving birth to immature and helpless animals that require care after birth for some time. Compare PRECOCIAL STRATEGY.

AMYLASE An enzyme that accelerates the hydrolysis (breaking down) of starch or glycogen.

ANALOGOUS STRUCTURES Parts of the anatomy with similar form and function and not based on descent from a common ancestor. Compare HOMOLOGOUS STRUCTURES.

ANTHROPOID Refers to the higher primates belonging to the suborder Anthropoidea (New and Old World monkeys, apes, and humans).

ANTIGEN A substance that produces an antibody response.

ANTISERUM A serum that contains antibodies that destroy a specific virus or bacterium.

APICAL FORAMEN The small opening at the root tip of a tooth through which blood vessels and nerves pass to the interior of the tooth.

ARBOREAL Refers to a tree-dwelling animal.

ARCHIPALLIUM The older part of the cortex of the cerebral hemispheres concerned primarily with smell. Compare NEOPALLIUM.

ASSOCIATION CORTEX (AREA) Those parts of the outer layers of the cerebral cortex concerned with sensory or motor functions.

AUDITORY BULLA The bone surrounding the middle-ear cavity and formed by the petrosal bone in primates.

AUTOCHTHONOUS Indigenous to a given ecosystem.

AUTOSOME Any chromosome other than the sex chromosomes.

BACULUM A small rod-shaped bone in the penis of many nonhuman primates.

BICORNUTE UTERUS A uterus consisting of a short median body surmounted by two conical horns which receive the uterian tubes.

BICUSPID A tooth possessing two cusps as do the premolars of many primates.

BILOPHODONT The molar teeth of Old World monkeys; the tooth has the two mesial (front) and the two distal (rear) cusps connected by transverse enamel lophs (ridges).

BINOCULAR VISION Vision in which the two visual axes focus on a distant object; a steroscopic (three-dimensional) image.

BINOMIAL SYSTEM The two-term system of classification established by Linnaeus. Every organism is identified by two Latin words, the genus name and, second, the specific name.

BIPEDAL A type of locomotion using the hindlimbs.

BRACHIATION A type of locomotion using only the forelimbs to swing beneath the branches of trees.

CANINIFORM Shaped like a canine tooth.

CATARRHINE All members of the infraorder Catarrhini, or Old World anthropoids (monkeys, apes, and humans).

CATHEMERAL Active, at various times, throughout the day and night.

CEMENTUM A hard calcified substance arranged in layers around the tooth root. It is a modified form of bone.

CENOZOIC ERA (65 MYA—present) The era following the Mesozoic era.

CHEEK POUCHES Outpouchings of the cheeks on each side of the face opposite the premolar teeth for the temporary storage of food; found only in the Cercopithecinae; also known as buccal pouches.

CHROMOSOMES Structures found in the nucleus which contain the hereditary material (DNA).

CLASS A category, or taxonomic rank, in the classification of animals or plants which includes those organisms that have adapted to a similar way of life (e.g., the class Mammalia).

CLAVICLE The bone connecting the sternum with the scapula and functioning as a strut; also known as the collarbone.

CORE AREA An area within the home range of an animal that is regularly used.

CORONOLATERAL SULCUS A sulcus (groove) passing longitudinally along the lateral surface of the cerebrum of many prosimians separating the representations of the head and forelimbs.

CRANIUM The skull, without the mandible (lower jaw).

CREPUSCULAR Refers to animals that are active at dawn and twilight.

CRISTA OBLIQUA An enamel crest on the upper molars of hominoids connecting the protocone and metacone.

CYTOARCHITECTURE The arrangement, orientation, density, and coloring features of nerve cells within the brain.

DECIDUOUS TEETH Often referred to as primary, baby, or milk teeth, these are shed and replaced by the permanent teeth. They are usually smaller than the permanent teeth that replace them.

DENTAL AGE Refers to the age of an organism based on the development of its dentition, for example, the amount of dental calcification present or the stage of dental eruption (see SKELETAL AGE).

DENTAL COMB The lower incisors and canines in Lemuridae, Indriidae, and Lorisidae projecting nearly straight forward (procumbent) from the front of the jaw and forming a comb used for grooming and scooping gums and resins from trees.

DENTAL FORMULA The number and types of teeth present in each quadrant of the upper and lower jaws. Old World monkeys, apes, and humans have the formula 2I-1C-2P-3M/2I-1C-2P-3M.

DENTINE A calcareous material similar to but harder and denser than bone and composing the principal mass of a tooth; covered by enamel on the crown and cementum on the root; forms the ivory in an elephant's tusk.

DERIVED CHARACTER STATE A later evolutionary character relative to its ancestral state; also called an apomorphy.

DIASTEMA (PL. DIASTEMATA) The space or gap between adjacent teeth (e.g., the space between the upper lateral incisor and canine in many monkeys and apes).

DIGITIGRADE Refers to the type of locomotion in which an animal supports its weight on its phalanges; compare palmigrade locomotion.

DIPHYLETIC TAXON A phylogenetic group that has evolved from two separate groups.

DIPLOID Possessing the full complement of chromosomes ($2n$) (46 in humans).

DIURNAL Refers to animals that are active during the day.

DNA (DEOXYRIBONUCLEIC ACID) A large molecule that carries the genetic code.

DRYOPITHECUS DENTAL PATTERN One of several occlusal patterns found on the lower molars of fossil and living hominoids; also known as the Y-5 pattern.

ECOLOGICAL NICHE The environment in which a species lives.

ECTOTYMPANIC BONE A ring of bone around the outside of the middle-ear chamber that supports the ear drum; also known as the tympanic ring or bone.

EMBRYO The name applied to organisms during their early stage of growth and development, a time period known as the embryonic period.

ENAMEL The hard, highly mineralized substance that covers the outer regions on the tooth crown; the hardest substance in an animal's body.

ENTEPICONDYLAR FORAMEN A hole in the distal (lower) end of the humerus in some primates through which the brachial artery and median nerve pass.

EOCENE EPOCH (55–35 MYA) The second epoch of the Cenozoic era, a time of many prosimian radiations.

ESTRUS The period when a female nonhuman primate ovulates and is receptive to copulation; also referred to as "heat."

EUPRIMATES A term often applied to "primates of modern aspect" (i.e., after the plesiadapiformes).

FAMILY A taxonomic rank that includes genera and species (e.g., the family Cercopithecidae). Family names always end with the suffix "idae."

FAUNA All the animal life in a given geographic region or period of time.

FAUNIVORES Animals that feed primarily on animal matter.

FETUS The unborn offspring of mammals in the later stages of development.

FLORA All the plant life in a given geographic region or period of time.

FOLIVORES Animals that feed primarily on leaves.

FRUGIVORES Animals that feed primarily on fruits.

GAMETE A germ cell, either male (sperm) or female (ovum), carrying a haploid or single set of chromosomes.

GENUS (PL. GENERA) The taxonomic rank larger than species and smaller than family. A genus usually includes a number of species.

GESTATION The period of pregnancy during which the embryo and fetus grow and develop.

GONDWANALAND One of the two supercontinents that broke away as Pangaea split apart about 200 MYA. Gondwanaland later separated into Africa, Antarctica, Arabia, Australia, the East Indies, India, Madagascar, and New Guinea. The other supercontinent was Laurasia.

GRANDE COUPURE A time at the end of the Eocene when many groups of land mammals became extinct, including primates.

GRAMNIVORES Animals that feed primarily on seeds.

GUMMIVORES Animals that feed primarily on gums and resins.

HALLUX The first digit on the foot; the "big toe."

HAPLOID The number of chromosomes (a single set or n) in the gametes after meiosis.

HAPLORHINI The suborder of primates that includes the tarsiers, New World and Old World monkeys, apes, and humans.

HERBIVORES Animals that feed primarily on vegetable matter.

HETEROCHRONY Refers to the evolutionary changes in timing, and or rate, of development.

HETERODONT Differentiation of the teeth for different functions, as incisors, canines, premolars, and molars. Compare HOMODONT.

HETEROMORPHIC Of different size and shape.

HOME RANGE The area in which an animal lives.

HOMINIDAE The family that includes all fossil and living members of the genera *Australopithecus* and *Homo*.

HOMODONT Having teeth of similar size and shape. Compare HETERODONT.

HOMOLOGOUS STRUCTURES Similarities in the anatomical structure of an organ or part in one kind of animal with the comparable structure in another due to common descent.

INSECTIVORES Animals that feed primarily on insects.

INTERMEMBRAL INDEX The ratio between the length of the forelimb (humerus + radius) divided by the length of the hindlimb (femur + tibia) × 100.

ISCHIAL CALLOSITIES Calloused areas (sitting pads) overlying the ischial tuberosities on Old World monkeys and gibbons.

ISCHIAL TUBEROSITIES A large prominence on the lower surface of the ischium supporting the ischial callosities in Old World monkeys and gibbons.

KARYOTYPE The chromosomes of a cell arranged according to size and banding pattern.

KAY'S THRESHOLD The approximate body weight (about 500 g) that separates predominately insectivorous from noninsectivorous primates.

KNUCKLE-WALKING Quadrupedal locomotion in which some of the

weight is placed on the knuckles (middle phalanges) as the animal walks; characteristic of chimpanzees and gorillas.

LAURASIA One of the two supercontinents resulting from the breakup of Pangaea about 200 MYA. It later separated into Asia, Europe, and North America.

MAMMARY GLANDS The milk glands of mammals; breasts.

MANUS The hand.

MEIOSIS The cell division that produces the haploid complement (n) of chromosomes; also known as reduction and division. Compare MITOSIS.

MIOCENE EPOCH (24–5 MYA) The fourth geological epoch of the Cenozoic era.

MITOSIS Cell division that produces two new cells identical to the original, with the same number of chromosomes. Compare MEIOSIS.

MOLARIFORM Molarlike in form and function.

MONOPHYLETIC A species containing all the known descendants of an ancestral species.

MORPHOLOGY The structure or form of an organism.

MOSAIC EVOLUTION The evolution of different characters at different rates.

NEONATE A newborn infant.

NEOPALLIUM The phylogenetically more recent part of the cerebrum overlying the archipallium; also known as the neocortex.

NOCTURNAL Refers to animals that are active at night.

NULLIPAROUS Refers to a female that has never given birth.

NOYAU A social structure in which the home ranges of the adults are separate but those of different sexes do overlap.

ODONTOLOGY The scientific study of animal teeth.

OLFACTORY Refers to the sense of smell.

OLIGOCENE EPOCH (35–24 MYA) The third geological epoch of the Cenozoic era.

OMNIVORE An animal that does not specialize in one type of food; it eats both meat and plants.

ONTOGENTIC Refers to the growth and development of organisms.

OPPOSABILITY The ability to rotate a digit (e.g., thumb) so that it comes in opposition with the other digits.

ORDER A taxonomic classification in the Animal Kingdom between class and genus (e.g., the order Primates).

OROGENY The process or processes of mountain formation.

ORTHOGRADE Walking with the body upright or vertical.

OSTEOLOGY The study of bones.

PANGAEA Former supercontinent composed of all the continental crust of the world and later fragmented by drift into Laurasia and Gondwanaland.

PARALLEL EVOLUTION Evolution of similar characteristics in different groups of organisms that were once related.

PECTORAL GIRDLE The system of bones (scapula and clavicle) supporting the forelimb in vertebrates. Also known as the shoulder girdle.

PENTADACTYLA Having five digits on the hand or foot.

PEPSIN A digestive enzyme found in the gastric juices of many vertebrates.

PERINEUM The region between the anus and the scrotum in the male, between the anus and the vulva in the female.

PES The foot

PHEROMONE Any substance secreted by an animal that influences the behavior of other individuals of the same species.

PHILTRUM The vertical furrow or cleft in the middle of the upper lip. Interestingly, both the Latin and Greek words mean a love charm or potion.

PLATE TECTONICS Global tectonics based on a model that explains the movements of the outer solid portions of the earth (lithosphere or plates) on the viscous underlying mantle (asthenosphere). Each plate moves more or less independently of the other.

PNEUMATIZATION The process which results in the formation of the bony air sinuses of the skull (e.g., the frontal sinuses in the frontal bone).

POLLEX The first digit of the manus; the thumb.

POLYMORPHIC Having a variety of forms.

PONGIDAE The family including the great apes, chimpanzees, gorillas, and orangutans.

POSTCRANIAL SKELETON All the skeleton except the skull.

POSTORBITAL BAR Bone enclosing the lateral side of the eye orbit in most fossil and living primates.

PRECOCIAL STRATEGY Animals born in an advanced (mature) stage of development and requiring little care. Compare ALTRICIAL.

PRONATION Medial rotation of the forearm toward the thumb side so that the palm faces backward or downward. Compare SUPINATION.

PROSIMIAN A lower primate belonging to the suborder Prosimii, including lemurs, lorises, galagos, and tarsiers.

PTERION The region on the lateral (temporal) side of the skull, that is, the junction of the temporal, sphenoid, parietal, and frontal bones.

PULP (DENTAL) The substance that occupies the central portion (pulp cavity) of the tooth and is surrounded by dentine. It houses the nerves and blood vessels supplying the tooth.

QUADRUPEDALISM Locomotion employing all four limbs.

RH-FACTOR An incompatability between the mother and fetus; also known as hemolytic disease of the newborn. First detected by using the sera of rabbits and guinea pigs immunized with the red cells of the rhesus monkey, hence the name Rh-factor.

RHINARIUM The area of specialized skin at the apex of the muzzle. It is hairless, smooth, and moist; present in strepsirhines.

SACCULATED STOMACH Found only in the folivorous primates, particularly in the colobines; it is greatly enlarged and has several chambers which help to digest the large amounts of leaves these animals consume.

SALIVARY GLANDS These glands secrete amylase, important in the digestion of starch. The major ones consist of the parotid, submandibular, and the greater and lesser sublingual glands.

SECTORIAL PREMOLAR The lower third premolar of Old World monkeys and pongids, enlongated and compressed from side to side to form a honing surface for the upper canine.

SEXUAL DIMORPHISM Differences in the size and shape of certain characteristics between males and females of the same species (e.g., body size and canine size).

SIMIAN SHELF A bony shelf on the inside of the mandibular symphysis, present in some fossil and in many living primates.

SIMIAN SULCUS A furrow on the surface of the cerebrum forming a boundry to the visual cortex; also known as the lunate sulcus.

SKELETAL AGE An estimate of age based upon the sequence and time of appearance and union of the various elements of the skeleton. Compare DENTAL AGE.

SPECIES A taxonomic rank below the genus composed of organisms living in the same region that can, or actually do, interbreed and produce fertile offspring.

SPLANCHOCRANIUM Portions of the skull derived from the primitive skeleton of the gill apparatus, that is, the bony face.

STREPSIRHINI The suborder of primates including the lemurs and lorises.

SUPINATION The rotation of the forearm so that the palm faces forward or upward. Compare PRONATION.

SUTURE A distinguishable line of union between two bones of the skull (e.g., the sagittal suture between the parietal bones).

SYMPATRIC SPECIES Two species that share the same geographical region. Compare ALLOPATRIC SPECIES.

TACTILE PADS Elevations on the palmar and plantar surfaces of the hands and feet bearing ridges and grooves that prevent slipping between the palmar (or plantar) skin and the substrate; also known as touch pads.

TAXON (PL. TAXA) A group of organisms, at any rank, that are classified together.

TERRITORY The part of a home range that a social group uses exclusively and defends.

TETHYS SEA A sea that existed in the early Tertiary separating Eurasia from Africa. The remnant of the Tethys Sea today is the Mediterranean Sea.

TRITUBERCULAR A triangular-shaped tooth possessing three cusps.

TYPE SPECIMEN An individual specimen that serves as the basis for the original name and description of the species and for identifying all other individuals of the same species, also known as the holotype.

VIBRISSAE Specialized hairs (whiskers) with sensory nerve endings in the skin at their bases.

ZOOGEOGRAPHY The science that describes and explains the distribution of animals in space and time.

ZYGOTE An organism produced by the union of an egg and a sperm; the fertilized ovum.

⤳ Bibliography

Alberch, P., S. J. Gould, G. F. Oster, and D. B. Wake. 1979. Size and shape in ontogency and phylogeny. *Paleobiology* 5:296–317.

Allman, J. 1982. Reconstructing the evolution of the brain in primates through the use of comparative neurophysiological and neuroanatomical data. In *Primate Brain Evolution Methods and Concepts*, pp. 13–28. Ed. E. Armstrong and D. Falk. Plenum Press, New York.

Anapol, F., and J. G. Fleagle. 1988. Fossil platyrrine forelimb bones from the early Miocene of Argentina. *Am. J. Phys. Anthropol.* 76:417–28.

Andrews, P., and J. Van Couvering. 1980. Paleoenvironments in the East African Miocene. In *Approaches to Primate Paleobiology*, pp. 62–103. Ed. F. Szalay. Karger, Basel.

Anemone, R. L., M. P. Mooney, and M. I. Siegel. 1996. Longitudinal study of dental development in chimpanzees of known chronological age: Implications for understanding the age at death of Plio-Pleistocene hominids. *Am. J. Phys. Anthrop.* 99:119–33.

Ankel-Simons, F. 1983. *A Survey of Living Primates and Their Anatomy.* Macmillan, New York.

———. 1996. Deciduous dentition of the aye-aye, *Daubentonia madagascariensis. Am. J. Primatol.* 39:87–97.

Arambulo, P., III, F. Encarnacion, J. Estupinian, and H. Samame, C. R. Watson, and R. R. Weeler, eds. 1993. *Primates of the Americas: Strategies for Conservation and Sustained Use in Biomedical Research.* Battelle Press, Columbus.

Armstrong, E. 1982. Mosaic evolution in the primate brain: Differences in the hominoid thalamus. In *Primate Brain Evolution Methods and Concepts*, pp. 131–62. Ed. E. Armstrong and D. Falk. Plenum Press, New York.

Ashley-Montagu, M. F. 1935. The Premaxilla in the primates. *Quart. Rev. Biol.* 10:32–59, 181–208.

Ashley-Montagu, M. F. 1943. *Edward Tyson, M.D., F.R.S. 1650–1708, and the Rise of Human and Comparative Anatomy in England.* Am. Philos. Soc. Philadelphia.

Badgely, C., Q. Guoqin, C. Wanyong, and H. Defen. 1988. Paleoecology of a Miocene, tropical, upland fauna: Lufeng, China. *Nat. Geograph. Res.* 4:178–95.

Beard, K. C. 1990. Gliding behaviour and paleoecology of the alleged primate family Paromomyidae (Mammalia, Dermoptera). *Nature* 345:340–41.

Beard, K. C., Tao Qi, M. R. Dawson, B. Wang, and C. Li. 1994. A diverse new primate fauna from Middle Eocene fissure-fillings in southeastern China. *Nature* 368:604–9.

Beard, K. C., Y. Tong, M. R. Dawson, J. Wang, and X. Huang. 1996. Earliest complete dentition of an anthropoid primate from the late Middle Eocene of Shanxi Province, China. *Science* 272:82–85.

Beck, B. B. 1980. *Animal Tool Behavior.* Garland Press, New York.

Benefit, B. R. 1990. Fossil evidence for the dietary evolution of Old World monkeys. *Am. J. Phys. Anthrop.* 81:193.

Benefit, B. R., and M. C. McCrossin. 1990. Diet, species diversity and distribution of African fossil baboons. *Kroeber Anthropol. Soc. Papers*, Nos. 71–72:71–93.

Benirschke, K., and M. H. Bogart. 1976. Chromosomes of the tan-handed titi (*Callicebus torquatus* Hoffmannsegg, 1807). *Folia Primatol.* 25:25–34.

Biegert, J. 1963. The evaluation of characteristics of the skull, hands, and feet for primate taxonomy. In *Classification and Human Evolution*, pp. 116–45. Ed. S. L. Washburn. Aldine, Chicago.

Blaney, S. P. A. 1986. An allometric study of the frontal sinus in *Gorilla, Pan* and *Pongo. Folia Primatol.* 47:81–96.

Bogin, B. 1988. *Patterns of Human Growth.* Cambridge University Press, Cambridge.
———. 1990. The evolution of human childhood. *Bioscience* 40:16–25.

Boulengener, Edward G. 1937. *Apes and Monkeys.* Robert M. McBride and Co., New York.

Bown, T. H., M. J. Kraus, S. Wing, B. Tiffiny, E. L. Simons, and C. F. Vondraw. 1982. The Fayum primate forest revisited. *J. Hum. Evol.* 11:603–32.

Brett, F., C. J. Jolly, W. W. Socha, and A. S. Wiener. 1976. Humanlike ABO blood groups in wild Ethiopian baboons. *Yearbook Phys. Anthrop.* 20:276–89.

Brizzee, K. R., and W. P. Dunlap. 1986. Growth. In *Reproduction and Development, Comparative Primate Biology*, Vol. 3, pp. 363–413. Ed. W. R. Dukelow and J. Erwin. Alan R. Liss, Inc., New York.

Buetter-Janusch, J., and R. J. Andrew. 1962. The use of the incisors by primates in grooming. *Am. J. Phys. Anthropol.* 20:127–30.

Cachel, S. 1979. A functional analysis of the primate masticatory system and the origin of the anthropoid post-orbital septum. *Am. J. Phys. Anthropol.* 50:1–18.
———. 1981. Plate tectonics and the problem of anthropod origins. *Yearbook Phys. Anthrop.* 24:139–72.

Carpenter, C. R. 1934. A field study of the behavior and social relations of the howling monkeys (*Alouatta palliata*). *Comp. Psychol. Monogr.* 10:3–92.
———. 1940. A field study in Siam of the behavior and social relations of the Gibbon (*Hylobates lar*). *Comp. Psychol. Monogr.* 16:1–212.
———. 1964. *Naturalistic Behavior of Nonhuman Primates.* Pennsylvania State University Press, University Park.

Cartmill, M. 1968. Morphology and orientation of the orbit in arboreal mammals. *Am. J. Phys. Anthropol.* 29:131–32.

———. 1972. Arboreal adaptations and the origin of the order Primates. In *Functional and Evolutionary Biology of Primates*, pp. 97–122. Ed. R. H. Tuttle, Aldine-Atherton, Chicago.

———. 1974. Pads and claws in arboreal locomotion. In *Primate Locomotion*, pp. 45–84. Ed. F. A. Jenkins, Jr. Academic Press, New York.

———. 1994. Nonhuman primates. In *The Cambridge Encyclopedia of Human Evolution*, pp. 24–32. Ed. S. Jones, R. Martin, and D. Pilbeam. Cambridge University Press, Cambridge.

Cartmill, M., and A. D. Yoder, 1994. Molecules versus morphology in primate systematics: An introduction. *Am. J. Phys. Anthropol.* 94:1.

Charles-Dominique, P. 1983. Ecology and social adaptation in didelphid marsupials: Comparison with eutherians of similar ecology. In *Advances in the Study of Mammalian Behavior*, pp. 395–422. Ed. J. F. Eisenberg and D. Kleiman, Special Publication No. 7, Am. Soc. Mammalogists.

Cheney, D. L., and R. W. Wrangham. 1987. Predation. In *Primate Societies*, pp. 227–39. Ed. B. B. Smuts, D. L. Cheney, R. M. Seyfarth, R. W. Wrangham, and T. T. Struhsaker. University of Chicago Press, Chicago.

Chiarelli, A. B., A. L. Koen, and G. Ardito. 1979. *Comparative Karyology of Primates.* Mouton Publishers, Paris.

Chivers, D. J. 1987. Conservation of gorillas and their habitats. *Internat. Primatol. Soc. Newsletter,* 10:3.

Ciochon, R., and A. B. Chiarelli. 1980. Paleobiogeographic perspectives on the origin of the Platyrrhini. In *Evolutionary Biology of the New World Monkeys and Continental Drift*, pp. 459–93. Ed. R. Ciochon and A. B. Chiarelli. Plenum Press, New York.

Ciochon, R., D. Savage, T. Tint, and B. Maw. 1985. Anthropoid origins in Asia? New discovery of *Amphipithecus* from the Eocene of Burma. *Science* 229: 756–59.

Clemens, W. 1974. *Purgatorius*, an early paromomyid primate. *Science* 184:903–5.

Coelho, A. M., Jr. 1985. Baboon dimorphism: Growth in weight, length, and adiposity from birth to 8 years of age. In *Nonhuman Primate Models for Human Growth*, pp. 125–59. Ed. E. S. Watts. Alan R. Liss, New York.

Cohen, J. E. 1995. Population growth and earth's human carrying capacity. *Science* 269:341–46.

Colbert, E. 1937. A new primate from the Upper Eocene Pondaung formations of Burma. *Am. Mus. Nat. Hist. Novit.* 951:1–18.

Conroy, G. 1976. Primate postcranial remains from the Oligocene of Egypt. *Contrib. Primatol.* 8:1–134.

———. 1980. Ontogeny, auditory structure and primate evolution. *Am. J. Phys. Anthropol.* 52:443–51.

———. 1987. Problems in body weight estimation in fossil primates. *Int. J. Primatol.* 8:115–37.

———. 1990. *Primate Evolution.* W. W. Norton and Co., New York.

Conroy, G., and K. Kuykendall. 1995. Paleopediatrics: Or when did human infants really become human? *Am. J. Phys. Anthropol.* 98:121–31.

Crockett, C. M., and J. F. Eisenberg. 1987. Howlers: Variations in group size and demography. In *Primate Societies*, pp. 54–68. Ed. B. B. Smuts, D. L. Cheney, R. M. Seyfarth, R. W. Wrangham, and T. T. Struhsaker. University of Chicago Press, Chicago.

Culotta, E. 1992. A new take on anthropoid origins. *Science* 256:1516–17.

———. 1995. Many suspects to blame in Madagascar extinctions. *Science* 268:1568–69.

Darwin, C. 1859. *The Origin of Species*. Murray, London.

———. 1872. *The Expression of the Emotions in Man and Animals*. Murray, London.

Deegan, J. F., II, and G. H. Jacobs. 1994. Photopigments of bush babies: Relationships to retinal organization and comparative features. *Am. J. Primatol.* 33:205.

Delson, E. 1994. Evolution of Old World monkeys. In *The Cambridge Encycopedia of Human Evolution*, pp. 217–22. Ed. S. Jones, R. Martin, and D. Pilbeam. Cambridge University Press, Cambridge.

DeVore, I., and K. R. L. Hall. 1965. Baboon ecology. In *Primate Behavior: Field Studies of Monkeys and Apes*, pp. 20–52. Ed. I. DeVore. Holt, Rinehart and Winston, New York.

DeVore, I., and S. L. Washburn. 1963. Baboon ecology and human evolution. In *African Ecology and Human Evolution*, pp. 335–67. Ed. F. C. Howell and F. Bourliere. Viking Fund Publications in Anthropology No. 36, Wenner-Gren Foundation, New York.

Disotell, T. R. 1994. Generic level relationships of the Papionini (Cercopithecoidea). *Am. J. Phys. Anthropol.* 94:47–57.

Du Chaillu, P. 1890. *Adventures in the Great Forest of Equatorial Africa and the Country of the Dwarfs*. Harper and Brothers, New York.

Duchin, L. E. 1990. The evolution of articulate speech: comparative anatomy of the oral cavity in *Pan* and *Homo*. *J. Hum. Evol.* 19:687–97.

Dutrillaux, B., M. Muleris, and J. Couturier. 1986. Chromosomal evolution of Cercopithecinae. In *A Primate Radiation: Evolutionary Biology of the African Guenons*, pp. 156–70. Ed. A. Gautier-Hion, F. Bourliere, and J. P. Gautier. Cambridge University Press, Cambridge.

Eccles, J. C. 1984. *The Human Mystery*. The Gifford Lectures, University of Edinburgh, 1977–78. Routledge and Kegan Paul, London.

Eimerl, S., and I. De Vore. 1977. *The Primates*. Time-Life Books, Alexandria.

Eisenberg, J. F., N. A. Muckenhirn, and R. Rudran. 1972. The relationship between ecology and social structure in primates. *Science* 176:863–74.

El Mahdy, Christine. 1989. *Mummies, Myth and Magic in Ancient Egypt*. Thames and Hudson, New York.

Eldredge, N., and J. Cracraft. 1980. *Phylogenetic Patterns and the Evolutionary Process*. Columbia University Press, New York.

Elliot, D. G. 1912. *A Review of the Primates*. American Museum of Natural History, New York.

Emel, L. M., L. Levitch, and D. R. Swindler. 1982. Sexual dimorphism in skeletal development of the hand of *Macaca nemestrina*. *Am. J. Phys. Anthropol.* 57:185.

Emihani, C., S. Gaerther, and B. Lioz. 1972. Neogene sedimentation of the Blake Plateau and the emergence of the Central American isthmus. *Palaeogeograph., Palaeoclimat., Palaeoecol.* 11:1–10.

Few, R. 1991. *Macmillan Illustrated Animal Encylopedia.* Maxwell Macmillan International, New York.

Fischman, J. 1995. Why mammal ears went on the move. *Science* 270:1436.

Fleagle, J. G. 1985. New primate fossils from Colhuehuapian deposits at Gaiman and Sacanana, Chubut Province, Argentina. *Armeghinians* 21:266–74.

———. 1988. *Primate Adaptation and Evolution.* Academic Press, London.

Fleagle, J. G., and R. A. Mittermeier. 1980. Locomotor behavior, body size and comparative ecology of seven Surinam monkeys. *Am. J. Phys. Anthropol.* 52: 301–22.

Fleagle, J. G., D. W. Powers, G. C. Conroy, and J. P. Watters. 1987. New fossil Platyrrhines from Santa Cruz Province, Argentina. *Folia Primatol.* 48:65–77.

Fleagle, J. G., and R. F. Kay. 1987. The phyletic position of the Parapithecidae. *J. Hum. Evol.* 16:483–532.

Fleagle, J. G., and R. F. Kay, eds. 1994. *Anthropoid Origins.* Plenum Press, New York.

Fooden, J. 1971. Male external genitalia and systematic relationships of the Japanese Macaque (*Macaca fuscata* Blyth, 1875). *Primates* 12:305–11.

Fowler, J. 1995. Education: The key to understanding. *The Explorers J.* 73:8–9.

Friedman, H. 1967. Colour vision in the Virginia opossum. *Nature* 213:835–36.

Galdikas, B. M. F. 1982. Orangutan Tool-Use at Tan Jung Reserve, Central Indonesian Borneo (Kalimantan Tengah). *J. Hum. Evol.* 11:19–33.

Gardner, R. A., and N. T. Gardner. 1969. Teaching sign language to a chimpanzee. *Science* 166:664–72.

Gavan, J. A. 1953. Growth and development of the chimpanzee: A longitudinal and comparative study. *Hum. Biol.* 25:93–143.

———. 1975. A classification of the order Primates. Museum Brief, No. 16. University of Missouri, Columbia.

Gavan, J. A., and D. R. Swindler. 1966. Growth rates and phylogeny in primates. *Am. J. Phys. Anthropol.* 24:181–90.

Gebo, D. L., and C. A. Chapman. 1995. Positional behavior in five sympatric Old World monkeys. *Am. J. Phys. Anthropol.* 97:49–76.

Geissmann, T., and M. Orgeldinger. 1995. Neonatal weight in gibbons (*Hylobates* spp.). *Am. J. Primatol.* 37:179–89.

Gibbons, A. 1990. Our chimp cousins get that much closer. *Science* 250:376.

———. 1994. Primate origins: New skull fuels debate. *Science* 266:541.

Gingerich, P. 1976. Cranial anatomy and evolution of early Tertiary Pesiadapidae (Mammalia, Primates). *Papers on Paleontology* No. 15. University of Michigan, Ann Arbor.

———. 1980. Dental and cranial adaptations in Eocene Adapidae. *Z. Morph. Anthropol.* 75:135–42.

———. 1981. Early Cenozoic Omomydiae and the evolutionary history of the Tarsiiform primates. *J. Hum. Evol.* 10:345–74.

———. 1984. Paleobiology of Tarsiiform primates. In *Biology of Tarsiers*, pp. 33–44. Ed. C. Niemitz. Gustav Fischer, Stuttgart.

————. 1984. Primate evolution. In *Univ. of Tennessee Studies in Geology*, 8 : 167 – 81. Ed. T. W. Broadhead. University of Tennessee, Knoxville.

————. 1986. Plesiadapis and delineation of the order Primates. In *Major Topics in Primate and Human Evolution*, pp. 31 – 46. Ed. B. Wood, L. Martin, and P. Andrews. Cambridge University Press, Cambridge.

————. 1989. New earliest Wasatchian mammalian fauna from the Eocene of northwestern Wyoming: Composition and diversity in a rarely sampled high-floodplain assemblage. *Papers on Paleontology* No. 28. University of Michigan, Ann Arbor.

Gingerich, P., B. Smith, and K. Rosenberger. 1982. Allometric scaling in the dentition of primates and prediction of body weight from tooth size in fossils. *Am. J. Phys. Anthropol.* 57 : 81 – 100.

Glaser, D. 1970. Über die Ossifikation de Extremitaten bein Neugenbornen Primaten (Mammalia). *Zeitschrift fur Morphologie der Tiere* 79 : 155 – 63.

————. 1972. Vergleichende Untersuchungen über den Geschmackssinn der Primaten. *Folia Primatol.* 17 : 267 – 72.

Glaser, D., H. Van der Wel, J. N. Brouwer, G. E. DuBois, and G. Hellekant. 1992. Gustatory responses in primates to the sweeter aspartame and their phylogenetic implications. *Chemical Senses* 17 : 325 – 35.

Goodall, J. 1970. Tool-using in primates and other vertebrates. In *Advances in the Study of Behavior* 3 : 165 – 87. Ed. D. Lehrman, R. Hinde, and E. Shaw. Academic Press, New York.

Goodall, J. 1986. *The Chimpanzees of Gombe*. The Belknap Press, Harvard University Press, Cambridge.

————. 1990. *Through a Window*. Houghton Mifflin, Boston.

Goodman, M. 1985. The rates of molecular evolution: The hominoid slowdown. *Bio Essays* 3 : 9 – 14.

Goodman, M., B. F. Koop, J. Czelusniak, D. H. A. Fitch, D. A. Tagle, and J. L. Slightom. 1989. Molecular phylogeny of the family of apes and humans. *Genome* 31 : 316 – 35.

Goodman, M., D. A. Tagle, D. H. A. Fitch, W. J. Bailey, and W. C. Czelusniak. 1990. Primate evolution at the DNA level and a classification of hominoids. *J. Mol. Evol.* 30 : 260 – 66.

Goodman, M., W. J. Bailey, K. Hayasaka, M. J. Stanhope, J. Slightom, and J. Czelusniak. 1994. Molecular evidence on primate phylogeny from DNA sequences. *Am. J. Phys. Anthropol.* 94 : 3 – 24.

Gould, S. J. 1977. *Ontogeny and Phylogeny*. Harvard University Press, Cambridge.

————. 1981. *The Mismeasure of Man*. W. W. Norton and Company, New York.

Gouzoules, S., and H. Gouzoules. 1987. Kinship. In *Primate Societies*, pp. 299 – 305. Ed. B. B. Smuts, D. L. Cheney, R. M. Seyfarth, R. W. Wrangham, and T. T. Strusaker. University of Chicago Press, Chicago.

Gregory, W. K. 1916. Studies on the evolution of the primates. *Bull. Am. Mus. Nat. Hist.* 35 : 239 – 355.

Gribbin, J. 1993. *In the Beginning, the Birth of the Living Universe*. Black Day Books, Little Brown, New York.

Gunnell, G. F. 1989. Evolutionary history of Microsyopoidea (Mammalia, ?Primates) and the relationship between Plesiadapiformes and primates. *Papers on Paleontology* No. 27. University of Michigan, Ann Arbor.

Gurche, J. 1982. Early primate brain evolution. In *Primate Brain Evolution: Methods and Concepts*, pp. 227–46. Ed. E. Armstrong and D. Falk. Plenum Press, New York.

Haines, D. E. 1986. The primate cerebellum. In *Systematics, Evolution, and Anatomy. Comparative Primate Biology*, 1:491–536. Ed. D. R. Swindler and J. Erwin. A. R. Liss, New York.

Haines, D. E., B. C. Albright, G. E. Goode, and H. M. Murray. 1974. The external morphology of the brain of some Lorisidae. In *Prosimian Biology*, pp. 673–724. Ed. R. D. Martin, G. A. Doyle, and A. C. Walker. Duckworth, London.

Hall, B. K. 1992. *Evolutionary Developmental Biology*. Chapman and Hall, New York.

Hall, K. R. L., and I. De Vore. 1965. Baboon social behavior. In *Primate Behavior*, pp. 53–110. Ed. I. De Vore. Holt Rinehart and Winston, New York.

Hamilton, W. S., J. D. Boyd, and H. W. Mossman. 1957. *Human Embryology*. W. Heffer and Sons, Cambridge.

Harrison, T. A. 1986. A reassessment of the phylogenetic relationships of Oreopithecus bambolii Gervais. *J. Hum. Evol.* 15:541–83.

Hartmann, Robert. 1901. *Anthropoid Apes*. D. Appleton and Co., New York.

Harvey, P. H., M. Kavanagh, and T. H. Clutton-Brock. 1978. Sexual dimorphism in primate teeth. *J. Zool. Lond.* 186:475–85.

Hayes, C. 1951. *The Ape in Our House*. Harper, New York.

Hendrickx, A. G., and M. L. Houston. 1971. Prenatal and postnatal development. In *Comparative Reproduction of Nonhuman Primates*, pp. 334–81. Ed. E. S. E. Hafez. C. C. Thomas, Springfield.

Hennig, W. 1966. *Phylogenetic Systematics*. University of Illinois Press, Urbana.

Hershkovitz, P. 1970a. The decorative chin. *Bull. Field Mus. Nat. Hist.* 41:6–10.

―――. 1970b. Cerebral fissural patterns in Platyrrhine monkeys. *Folia Primat.* 13:213–40.

―――. 1977. *Living New World Monkeys (Platyrrhini): With An Introduction to Primates.* Vol. 1. University of Chicago Press, Chicago.

Hill, W. C. O. 1957. Pharynx, oesophagus, stomach, small and large intestine form and position. In *Primatologia: Handbook of Primatology* 312:139–207. Ed. H. Hofer, A. H. Schultz, and D. Starck. S. Karger, New York.

―――. 1972. *Evolutionary Biology of the Primates*. Academic Press, London.

Ho, C. K. 1988. Human origins in Asia? *Hum. Evol.* 3:357–65.

Ho, C. K., G. X. Zhou, and D. R. Swindler. 1995. Dental evolution of the orangutan in China. *Hum. Evol.*, 10:249–64.

Hobson, C. 1987. *The World of the Pharaohs*. Thames and Hudson, New York.

Höffstetter, R. 1974. Phylogeny and geographical deployment of the primates. *J. Hum. Evol.* 3:327–50.

―――. 1980. Origin and deployment of New World monkeys emphasizing the southern continent's route. In *Evolutionary Biology of the New World Monkeys and Continental Drift.* pp. 103–22. Ed. R. Ciochon and A. B. Chiarelli. Plenum Press, New York.

Holloway, R. L., Jr. 1968. The evolution of the primate brain: Some aspects of quantitative relations. *Brain Res.* 7:121–72.

Hooijer, D. A. 1948. Prehistoric teeth of man and the orang-utan from Central Sumatra, with notes on the fossil orang-utan from Java and Southern China. *Zoolog.-Mededelingen* 29:173–301.

Hrdy, S. B. 1977. *The Langurs of Abu.* Harvard University Press, Cambridge.

Hrdy, S. B., and P. L. Whitten. 1987. Patterning of sexual activity. In *Primate Societies*, pp. 370–84. Eds. B. B. Smuts, D. L. Cheney, R. M. Seyfarth, R. W. Wrangham, and T. T. Strumsaker. University of Chicago Press, Chicago.

Itoigawa, N., Y. Sugiyama, G. P. Sackett, and R. K. R. Thompson, eds. 1992. *Topics in Primatology: Behavior, Ecology, and Conservation.* University of Tokyo Press, Tokyo.

Iwamoto, M. 1975. On a skull of a fossil macaque from the Shikimizu limestone quarry in the Shikoku district, Japan. *Primates* 16:83–94.

Jablonski, N. G. 1993. Quaternary environments and the evolution of primates in East Asia, with notes on two new specimens of fossil cercopithecidae from China. *Folia Primatol.* 60:118–32.

———. 1993. *Theropithecus: The Rise and Fall of a Primate Genus.* Cambridge University Press, Cambridge.

———. 1995. Primate life histories and primate conservation. In *Primate Research and Conservation*, pp. 113–17. Ed. X. Wuping and Y. Zhang. China Forestry Publishing House, Beijing.

Jacobs, G. H. 1995. Variations in primate color vision: Mechanisms and utility. *Evol. Anthropol.* 3:196–205.

Jacobs, G. H., and J. F. Deegan, II. 1993. Photopigments underlying color vision in ringtail lemurs (*Lemur catta*) and brown lemurs (*Eulemur fulvus*). *Am. J. Primatol.* 30:243–56.

Jacobs, G. H., J. Neitz, and M. Neitz. 1993b. Genetic basis of polymorphism in the color vision of platyrrhine monkeys. *Vis. Res.* 33:269–74.

Jacobs, G. H., J. F. Deegan, II, J. Neitz, and M. A. Crognale. 1993c. Photometrics and color vision in the nocturnal monkey, *Aotus. Vis. Res.* 33:1773–83.

Jacobsen, N. 1970. Salivary amylase II: Alphas amylase in salivary glands of the *Macaca irus* monkey, the *Cercopithecus aethiops* monkey and man. *Caries. Res.* 4:200–205.

Jay, P. 1965. The common langur of North India. In *Primate Behavior Field Studies of Monkeys and Apes*, pp. 197–249. Ed. I. De Vore. Holt, Rinehart and Winston, New York.

Jerison, H. J. 1982. Brain size, cortical surface, and convolutedness. In *Primate Brain Evolution Methods and Concepts*, pp. 77–84. Ed. E. Armstrong and D. Falk. Plenum Press, New York.

Jolly, A. 1995. President's corner. *Internat. Primatol. Soc. Newsletter* 22:1.

Jones, S., R. Martin, and D. Pilbeam, eds. *The Cambridge Encyclopedia of Human Evolution.* Cambridge University Press, Cambridge.

Kavanagh, M. 1984. *A Complete Guide to Monkeys, Apes and Other Primates.* Viking Press, New York.

Kay, R. F. 1975. The functional adaptations of primate molar teeth. *Am. J. Phys. Anthropol.* 43:195–216.

———. 1984. On the use of anatomical features to infer foraging behavior in extinct primates. In *Adaptations for Foraging in Nonhuman Primates*, pp. 21–53. Ed. J. Cant and P. Rodman. Columbia University Press, New York.

Kay, R. F., and W. L. Highlander. 1978. The dental structure of mammalian folivores with special reference to primate and Phalangeroidea (Marsupialia). In *The Ecology of Arboreal Folivores*, pp. 173–91. Ed. G. G. Montgomery. Smithsonian Institution Press, Washington, D.C.

Kay, R. F., C. Ross, and B. A. Williams. 1997. Anthropoid origins. *Science* 275:797–804.

Kay, R. F., R. W. Thorington, Jr., and P. Houde. 1990. Eocene Plesiadapiform shows affinities with flying lemurs not primates. *Nature* 345:342–44.

Kay, R. F., J. G. M. Thewissen, and A. D. Yoder. 1992. Cranial anatomy of *Igancious graybullianus* and the affinities of the Plesiadapiformes. *Am. J. Phys. Anthropol.* 89:477–98.

Kelley, J. 1994. Evolution of apes. In *The Cambridge Encyclopedia of Human Evolution*, pp. 223–30. Ed. S. Jones, R. Martin, and D. Pilbeam. Cambridge University Press, Cambridge.

Kelley, J., and D. Pilbeam. 1986. The Dryopithecines: Taxonomy, anatomy, and phylogeny of Miocene large hominoids. In *Comparative Primate Biology*, Vol. 1, pp. 361–411. Ed. D. R. Swindler and J. Erwin. Alan R. Liss, New York.

Kelley, J., and D. Etler. 1989. Hominoid dental variability and species number at the Late Miocene site Lufeng, China. *Am. J. Primatol.* 18:15–34.

King, B. F. 1986. Morphology of the placenta and fetal membranes. In *Comparative Primate Biology*, Vol. 3, pp. 311–31. Ed. W. R. Dukelow and J. Erwin. Alan R. Liss, New York.

Kingdon, J. 1971. *East African Mammals: An Atlas of Evolution in Africa*. Vol. 1. Academic Press, London.

Koppe, T., and Hiroshi, N. 1995. On the morphology of the maxillary sinus floor in old world monkeys: A study based on three-dimensional reconstructions of CT scans. *Proceedings of the 10th International Symposium on Dental Morphology*, pp. 423–27. Ed. Ralf J. Radlanski and Herbert Renz. C. & M. Brunne, Berlin.

Kraus, B. S., and R. E. Jordan. 1965. *The Human Dentition Before Birth*. Lea and Febiger, Philadelphia.

Landsteiner, K. 1901. Agglutinationserscheinungen normalen menschlichen Blutes. *Wien Klin. Wschr.* 14:1132–34.

Landsteiner, K., and A. S. Weiner. 1940. An agglutinable factor in human blood recognized by immune sera for rhesus blood. *Proc. Soc. Exp. Med.* 43:223–24.

Lapin, B., and E. Fridman. 1975. *Monkeys for Science*. Novosti Press Agency Publishing House.

Laporte, L. F., and A. L. Zihlman. 1983. Plates, climate and hominoid evolution. *S. Af. J. Sci.* 79:149–71.

Le Gros Clark, W. E. 1971. *The Antecedents of Man: An Introduction to the Evolution of the Primates* (3d ed.). Quadrangle Books, Chicago.

Lestrel, P. E., and J. E. Sirianni. 1982. The cranial base in *Macaca nemestrina*: Shape changes during adolescence. *Hum. Biol.* 54:7–15.

Leutenegger, W. 1974. Functional aspects of pelvic morphology in simian primates. *J. Hum. Evol.* 3:207–22.

Lewin, R. 1988. Living in the fast track makes for small brains. *Science* 242:513–14.

MacDonald, D. 1985. *All the World's Animals: Primates*. Torstar Books, New York.

MacDonald, J. 1963. The Miocene faunas from the Wounded Knee area of western South Dakota. *Bull. Am. Mus. Nat. Hist.* 125:139–238.

Marks, J. 1994. Chromosomal evolution in primates. In *The Cambridge Encyclopedia of Human Evolution*, pp. 298–302. Ed. S. Jones, R. D. Martin, and D. Pilbeam. Cambridge University Press, Cambridge.

Marshall, J., and J. Sugardjito. 1986. Gibbon systematics. In *Comparative Primate Biology* 1:137–86. Ed. D. R. Swindler and J. Erwin. Alan R. Liss, New York.

Martin, R. D. 1975. The bearing of reproductive behavior and ontogeny on Strepsirhine phylogeny. In *Phylogeny of the Primates*, pp. 265–97. Ed. W. P. Luckett and F. S. Szalay. Plenum Press, New York.

———. 1990. *Primate Origins and Evolution: A Phylogenetic Reconstruction*. Chapman and Hall, London.

———. 1993. Primate origins: Plugging the gaps. *Nature* 363:223–34.

———. 1994. Classification of primates. In *The Cambridge Encyclopedia of Human Evolution*, pp. 17–23. Ed. S. Jones, R. D. Martin, and D. Pilbeam. Cambridge University Press, Cambridge.

McGrew, W. C. 1979. Evolutionary implications of sex differences in chimpanzee predation and tool use. In *The Great Apes: Perspectives on Human Evolution*, Vol. 5. Ed. D. A. Hamburg and E. R. McCown. Benjamin/Cummings, Menlo Park, California.

McKenna, M. C. 1975. Toward a phylogenetic classification of the Mammalia. In *Phylogeny of the Primates*, pp. 21–46. Ed. W. Luckett and F. S. Szalay. Plenum Press, New York.

———. 1980. Early history and biogeography of South America's extinct land mammals. In *Evolutionary Biology of the New World Monkeys and Continental Drift*, pp. 43–78. Ed. R. Ciochon and A. B. Chiarelli. Plenum Press, New York.

McKinney, M. L. 1988. *Heterochrony in Evolution: A Multidisciplinary Approach*. Plenum Press, New York.

Mittermeier, R. A., and J. G. Fleagle. 1976. The locomotor and postural repertoires of *Ateles geoffroyi* and *Colobus guereza*, and a re-evaluation of the locomotor category semibrachiation. *Am. J. Phys. Anthropol.* 45:235–51.

Mittermeier, R. A., and D. L. Cheney. 1987. Conservation of primates and their habitats. In *Primate Societies*, pp. 477–90. Ed. B. B. Smuts, D. L. Cheney, R. M. Seyfarth, R. W. Wrangham, and T. T. Struhsaker. University of Chicago Press, Chicago.

Mittermeier, R. A., W. R. Konstant, M. E. Nicoll, and O. Langrand. 1992. *Lemurs of Madagascar: An Action Plan for Their Conservation*. IUCN, Gland, Switzerland.

Mivart, St. George. 1873. On *Lipilemur* and *Cheirogaleus* and on the zoological rank of the Lemuroidea. *Proc. Zool. Soc. Lond.*, pp. 484–510.

Miyamoto, M. M., and M. Goodman. 1990. DNA systematics and evolution of primates. *Annu. Rev. Ecol. Syst.* 21:197–220.

Morell, V. 1994. Will primate genetics split one gorilla into two? *Science* 265:1661.

Mouri, Toshio. 1984. A morphological analysis of the pterion in genus *Alouatta*. *Reports of New World Monkeys* 4:29–34.

Moya-Sola, S., and M. Kohler. 1996. A *Dryopithecus* skeleton and the origins of great-ape locomotion. *Nature* 379:156–59.

Myers, N. 1979. *The Sinking Ark*. W. W. Norton, New York.

Napier, J. R. 1970. *The Roots of Mankind*. Smithsonian Institution Press, Washington, D.C.

Napier, J. R., and P. H. Napier. 1967. *A Handbook of Living Primates: Morphology, Ecology and Behaviour of Nonhuman Primates*. Academic Press, London.

Napier, J. R., and R. H. Napier. 1985. *The Natural History of the Primates*. MIT Press, Cambridge.

Nelson, H., and R. Jurmain. 1991. *Introduction to Physical Anthropology*. West, New York.

Norton, B. G., M. Hutchins, E. F. Stevens, and T. L. Maple, eds. 1995. *Ethics of the Ark: Zoos, Animal Welfare, and Wildlife Conservation*. Smithsonian Institution Press, Washington, D.C.

Nsanjama, H. 1995. Mountain gorillas killed in Uganda. *Laboratory Primate Newsletter* 34:16.

Pagel, M. D., and P. H. Harvey. 1988. How mammals produce large-brained offspring. *Evolution* 42:948–59.

Pan Yuerong. 1988. Small fossil primates from Lufeng, a latest Miocene site in Yunnan Province, China. *J. Hum. Evol.* 17:359–66.

Pereira, M. E., and L. A. Fairbanks, eds. 1993. *Juvenile Primates: Life History, Development, and Behavior*. Oxford University Press, New York.

Peyer, B. 1968. *Comparative Odontology*. University of Chicago Press, Chicago.

Phillips-Conroy, J. E., and C. J. Jolly. 1988. Dental eruption schedules of wild and captive baboons. *Am. J. Primatol.* 15:17–29.

Pickford, M. 1983. Sequence and environments of the lower and middle Miocene hominoids of western Kenya. In *New Interpretations of Ape and Human Ancestry*, pp. 421–40. Ed. R. Ciochon and R. Corruccini. Plenum Press, New York.

Plavcan, J. M., and C. P. van Schaik. 1992. Intrasexual competition and canine dimorphism in anthropoid primates. *Am. J. Phys. Anthropol.* 87:461–78.

Pocock, R. I. 1918. On the external characters of the lemurs and of *Tarsius*. *Proc. Zool. Soc. Lond.* 1:19–53.

Porter, C. A., I. Sampaio, H. Schneider, M. P. C. Schneider, J. Czelusniak, and M. Goodman. 1995. Evidence on primate phylogeny from E. globin gene sequences and flanking regions. *J. Mol. Evol.* 40:30–55.

Radetsky, P. 1995. Gut. *Discover*, May, 1995, pp. 76–81.

Radinsky, L. 1975. Primate brain evolution. *Am. Sci.* 63:656–63.

———. 1982. Some cautionary notes on making inferences about relative brain size. In *Primate Brain Evolution: Methods and Concepts*, pp. 29–37. Ed. E. Armstrong and D. Falk. Plenum Press, New York.

Ranson, S. W., and S. L. Clark. 1953. *The Anatomy of the Nervous System: Its Development and Function.* W. B. Saunder, Philadelphia.

Rasmussen, D. T. 1986. Anthropoid origins: A possible solution to the Adapidae-Omomyidae paradox. *J. Hum. Evol.* 15:1–12.

Rawlins, R. G., and M. J. Kessler, eds. 1986. *The Cayo Santiago Macaques: History, Behavior, and Biology.* State University of New York Press, Albany.

Raymo, C. 1983. *The Crust of the Earth.* Prentice-Hall, New Jersey.

Rea, D. K., J. C. Zachos, R. M. Owen, and P. Gingerich. 1990. Global change at the Paleocene-Eocene boundary: Climatic and evolutionary consequences of tectonic events. *Palaeogeograph., Palaeoclimat., Palaeoecol.* 79:117–28.

Reynolds, V. 1967. *The Apes, the Gorilla, Chimpanzees, Orangutan and Gibbon: Their History and Their World.* E. P. Dutton and Co., New York.

Richard, A. F. 1985. *Primates in Nature.* W. H. Freeman and Co., New York.

———. 1987. Malagasy prosimians: Female dominance. In *Primate Societies*, pp. 25–33. Ed. B. B. Smuts, D. L. Cheney, R. M. Seyfarth, R. W. Wrangham, and T. T. Strusaker. University of Chicago Press, Chicago.

Robinson, P. 1968. The paleontology and geology of the Badwater Creek area, central Wyoming, Part 4. *Annals. Carnegie Mus.* 39:307–26.

Rose, K. D., and J. M. Rensberger. 1983. Upper dentition of *Ekgmowechashala* (Omomyid Primate) from the John Day Formation, Oligo-Miocene of Oregon. *Folia Primatol.* 41:102–11.

Rosenberg, K., and W. Trevathan. 1996. Bipedalism and human birth: The obstetrical dilemma revisited. *Evolutionary Anthropology* 4:161–68.

Rosenberger, A. L. 1994. Evolution of New World monkeys. In *The Cambridge Encyclopedia of Human Evolution*, pp. 209–16. Ed. S. Jones, R. D. Martin, and D. Pilbeam. Cambridge University Press, Cambridge.

Ross, C. 1995. Muscular and osseous anatomy of the primate anterior temporal fossa and the functions of the postorbital septum. *Am. J. Phys. Anthropol.* 98:275–306.

Rumpler, Y., S. Warter, M. Hauwy, B. Meier, A. Peyrieras, R. Albignac, J. J. Petter, and B. Dutrillaux. 1995. Cytogenetic study of *Allocebus trichotis*: A Malagasy prosimian. *Am. J. Primatol.* 36:239–44.

Sade, D. S. 1964. Seasonal cycle in size of testes of free-ranging *Macaca mulatta*. *Folia Primatol.* 2:171–80.

Sanderson, I. T. 1957. *The Monkey Kingdom: An Introduction to the Primates.* Hanover House, New York.

Sarich, V., and M. Cronin. 1976. Molecular systems of the primates. In *Molecular Anthropology*, pp. 141–70. Ed. M. Goodman and R. E. Tashian. Plenum Press, New York.

Savage, D., and D. Russell. 1983. *Mammalian Paleofaunas of the World.* Addison-Wesley, Reading, Mass.

Savage, T. S. 1847. Notice of the external characters and habits of Troglodytes gorilla, a new species of orang from the Gaboon River. With J. Wyman, Osteology of the same. *Boston J. Nat. Hist.* 5:417–43.

Schoeninger, M. J. 1976. Functional significance of the development of the mesostyle in the Eocene primates, *Pelycodus* and *Notharctus. Am. J. Phys. Anthropol.* 44:204.

Schultz, A. H. 1926. Fetal growth of man and other primates. *Quart. Rev. Biol.* 1:465–521.

———. 1937. Fetal growth and development of the rhesus monkey. *Contrib. Embryol.* 155:73–97. Carnegie Inst., Washington, D.C.

———. 1940. Growth and develoment of the chimpanzee. *Contrib. Embryol.* 28:1–63. Carnegie Inst., Washington, D.C.

———. 1956. Postembryonic age changes. In *Primatologia* 1:887–964. Karger, Basel.

———. 1961. Vertebral column and thorax. In *Primatologia* 4:1–66. Karger, Basel.

———. 1963. Age changes, sex differences, and variability as factors in the classification of primates. In *Classification and Human Evolution*, pp. 85–115. Ed. S. L. Washburn, Aldine, Chicago.

———. 1971. The rise of primatology in the twentieth century. *Proc. 3rd Int. Congr. Primat. Zurich* 1:2–15. Karger, Basel.

———. 1972. *The Life of Primates.* Universe Books, New York.

Schuman, E. L., and C. L. Brace. 1955. Metric and morphological variation in the dentition of the Liberian chimpanzee. *Hum. Biol.* 26:239–68.

Seyfarth, R. M. 1987. Vocal communication and its relation to language. In *Primate Societies*, pp. 440–51. Ed. B. B. Smuts, D. L. Cheney, R. M. Seyfarth, R. W. Wrangham, and T. T. Struhsaker. University of Chicago Press, Chicago.

Shea, B. T. 1981. Relative growth of the limbs and trunk in the African apes. *Am. J. Phys. Anthropol.* 56:179–202.

———. 1983. Allometry and heterochrony in the African apes. *Am. J. Phys. Anthropol.* 62:275–89.

———. 1984. Between the gorilla and the chimpanzee: A history of debate concerning the existence of the Kooloo-Kamba or gorillalike chimpanzee. *J. Ethnobiol.* 4:1–13.

———. 1989. Heterochrony in human evolution: The case for neoteny reconsidered. *Yearbook Phys. Anthropol.* 32:69–101.

Siebert, J. R., and D. R. Swindler. 1984. Dental arch form in the Cercopithecidae. *Primates* 25:507–18.

Simon, N., and P. Geroudet. 1970. *Last Survivors: The Natural History of Animals in Danger of Extinction.* World, New York.

Simons, E. L. 1990. Discovery of the oldest known anthropoidean skull from the Paleocene of Egypt. *Science* 247:1521–1612.

———. 1993. Lemurs in the wild: Discovery of the Western Aye-Aye. *Lemur News* 1:6.

———. 1995. Egyptian Oligocene primates: A review. *Yearbook Phys. Anthropol.* 38:199–238.

Simons, E. L., and A. Wood. 1968. Early Cenozoic mammalian faunas, Fayum Province, Egypt. *Peabody Mus. Nat. Hist. Yale Univ. Bull.* 28:1−105.

Simons, E. L., and T. M. Brown, 1985. *Afrotarsius chatrathi, first tarsiiform primate* (? Tarsiidae) from Africa. *Nature* 313:457−77.

Simons, E. L., and D. T. Rasmussen. 1994. A remarkable cranium of *Plesiopithecus teras* (Primates, Prosimii) from the Eocene of Egypt. *Proc. Natl. Acad. Sci.* 19:9946−50.

Simpson, G. G. 1962. *Principles of Animal Taxonomy*, pp. vii-247. Columbia University Press, New York.

Sirianni, J. E., A. L. Van Ness and D. R. Swindler, 1982. Growth of the mandible in adolescent pigtailed macaques (*Macaca nemestrina*). *Hum. Biol.* 54:31−43.

Sirianni, J. E., and D. R. Swindler. 1985. *Growth and Development of the Pigtailed Macaque.* CRC Press, Boca Raton.

Smuts, B. B. 1987. Gender, agression and influences. In *Primate Societies*, pp. 400−412. Ed. B. B. Smuts, D. L. Cheney, R. M. Seyfarth, R. W. Wrangham, and T. T. Strusaker. The University of Chicago Press, Chicago.

Smuts, B. B., D. L. Cheney, R. M. Seyfarth, R. W. Wrangham, and T. T. Struhsaker, eds. 1987. *Primate Societies.* University of Chicago Press, Chicago.

Sneath, P. H., and R. R. Sokal. 1973. *Numerical Taxonomy* (2d ed.). W. H. Freeman, New York.

Socha, W. W. 1986. Blood groups of nonhuman primates. In *Systematics, Evolution and Anatomy: Comparative Primate Biology*, Vol. 1, pp. 294−333. Ed. D. R. Swindler and J. Erwin. Alan R. Liss, New York.

Socha, W. W., A. S. Wiener, J. Moor-Jankowski, and C. J. Jolly. 1977. Blood groups of baboons: Population genetics and feral animals. *Am. J. Phys. Anthropol.* 47:435−42.

Socha, W. W., and J. Moor-Jankowski. 1979. Blood groups of anthropoid apes and their relationship to human groups. *J. Hum. Evol.* 8:453−65.

Socha, W. W., A. Blancher, and J. Moor-Jankowski. 1995. Red cell polymorphisms in nonhuman primates: A review. *J. Med. Primatol.* 24:282−305.

Southwick, C. H. 1963. *Primate Social Behavior.* Van Nostrand Reinhold, New York.

Steklis, H. D. 1995. Death in Rwanda. *Laboratory Primate Newsletter* 34:6.

Stephan, H. 1972. Evolution of primate brains: A comparative anatomical investigation. In *The Functional and Evolutionary Biology of Primates*, pp. 155−74. Ed. R. Tuttle. Aldine-Atherton, Chicago.

Stevens, J. L., V. R. Edgerton, D. E. Haines, and D. M. Meyer. 1981 *An Atlas and Source Book of the Lesser Bushbaby (Galago senegalensis).* CRC Press, Boca Raton.

Stewart, K. J., and A. H. Harcourt. 1987. Gorillas: Variations in female relationships. In *Primate Societies*, pp. 155−64. Ed. B. B. Smuts, D. L. Cheney, R. M. Seyfarth, R. W. Wrangham, and T. T. Strusaker. University of Chicago Press, Chicago.

Strier, K. B. 1994. Myth of the typical primate. *Yearbook Physical Anthropology* 37:233−71.

Sussman, R. W., and W. G. Kinzey. 1984. The ecological role of the Callitrichidae: A review. *Am. J. Phys. Anthropol.* 64.419−49.

Swindler, D. R. 1976. *Dentition of Living Primates.* Academic Press, London.

Swindler, D. R., T. W. Jenkins, and A. W. Weiss, Jr. 1968. Fetal growth and development. In Biology of the Howler Monkey (Aloutta caraya). Biol. Primat. 7:28–47. Karger, Basel.

Swindler, D. R., and C. D. Wood. 1973. An Atlas of Primate Gross Anatomy: Baboon, Chimpanzee, and Man. University of Washington Press, Seattle.

Swindler, D. R., and L. H. Tarrant. 1973. The topography of the premaxillary-frontal region in nonhuman primates. Folia Primatol. 19:18–23.

Swindler, D. R., and J. E. Sirianni. 1975. Dental size and dietary habits of primates. Yearbook Phys. Anthropol. 19:166–82.

Swindler, D. R., and L. M. Emel. 1990. Dental development, skeletal maturation and body weight at birth in pig-tail macaques (Macaca nemestrina). Archs. Oral Biol. 35:289–94.

Swindler, D. R., and D. Meekins. 1991. Dental development of the permanent mandibular teeth in the baboon, Papio cynocephalus. Am. J. Hum. Biol. 3:571–80.

Szalay, F. S. 1976. Systematics of the Omomydidae (Tarsiiformes Primates) taxonomy, phylogeny and adaptation. Bulletin Am. M. Nat. Hist., 156:157–450.

Tanner, J. M. 1981. A History of the Study of Human Growth. Cambridge University Press, Cambridge.

Tattersall, I. 1979. Patterns of activity in the Mayotte Lemur, Lemur fulvus mayotensis. J. Mammal. 60:314–23.

Teaford, M. F. 1994. Dental microwear and dental function. Evolutionary Anthropology 3:17–30.

Tuttle, R. H. 1969. Knuckle-walking and the problem of human origins. Science 166:953–61.

Van Valen, L. 1994. The origin of the plesiadapid primates and the nature of Purgatorius. Evolutionary Monographs 15:3–79.

Van Valen, L., and R. Sloan. 1965. The earliest primates. Science 150:743–45.

Van Wagenen, G., and H. R. Catchpole. 1956. Physical growth of the rhesus monkey (Macaca mulatta). Am. J. Phys. Anthropol. 14:245–73.

Van Wagenen, G., and C. W. Asking. 1964. Ossification in the fetal monkey (Macaca mulatta). Am. J. Anat. 114:107–32.

Walls, G. L. 1942. The Vertebrate Eye and Its Adaptive Radiation. Cranebrook Institute of Science, Bloomfield, Michigan.

Ward, S. C., and W. H. Kimbel. 1983. Subnasal alveolar morphology and the systematic position of Sivapithecus. Am. J. Phys. Anthropol. 61:157–71.

Watson, J. O., and F. H. C. Crick. 1953. The structure of DNA. Cold Spring Harbor Symp. Quant. Biol. 18:123–35.

Watts, E. S. 1986. Skeletal development. In Reproduction and Development: Comparative Primate Biology, Vol. 3, pp. 415–39. Ed. W. R. Dukelow and J. Erwin. Alan R. Liss, New York.

Watts, E. S., and J. A. Gavan. 1982. Postnatal growth of nonhuman primates: The problem of the adolescent spurt. Hum. Biol. 54:53–70.

Webster, D., and M. Webster. 1974. Comparative Vertebrate Morphology. Academic Books, London.

Weinmann, J. P., and H. Sicher. 1947. Bone and Bones: Fundamentals of Bone Biology. Mosby, St. Louis.

Wilson, E. O. 1992. *The Diversity of Life*. Harvard University Press, Cambridge.

Wilson, J. A. 1966. A new primate from the earliest Oligocene, West Texas, preliminary report. *Folia Primatol.* 4:227–48.

Winkler, L. 1986. Relationships between the facial cheek pads and facial anatomy in the orangutan. *Am. J. Phys. Anthropol.* 69:280.

Wolfe, R. G. 1984. New specimens of the primate *Branisella boliviana* from the Early Oligocene of Salla, Bolivia. *J. Vert. Paleo.* 4:570–74.

Wolfheim, J. H. 1983. *Primates of the World: Distribution, Abundance and Conservation*. University of Washington Press, Seattle.

Wyman, J. *See* Savage, T. S.

Yeager, C., L. Isbell, and T. Young. 1995. Primate behavior and conservation: Missing links? *Evol. Anthropol.* 3:191.

Yoder, A. D. 1994. Relative position of the Cheirogaleidae in Strepsirhine phylogeny: A comparison of morphological and molecular methods and results. *Am. J. Phys. Anthropol.* 94:25–26.

Yokoyama, S., and R. Yokoyama. 1989. Molecular evolution of human visual pigment genes. *Molec. Biol. Evol.* 6:186–97.

Young, J. Z. 1971. *An Introduction to the Study of Man*. Oxford University Press, Oxford.

Zeller, A. C. 1987. Communication by sight and smell. In *Primate Societies*, pp. 433–39. Ed. B. B. Smuts, D. L. Cheney, R. M. Seyfarth, R. W. Wrangham, and T. T. Struhsaker. The University of Chicago Press, Chicago.

Zimmer, C. 1995. Tooling through the trees, pp. 46–47. *Discover*, November.

Zingeser, M. R. 1970. The morphological basis of the underbite trait in Langurs (*P. melalophus, T. cristatus*) with an analysis of adaptive and evolutionary implications. *Am. J. Phys. Anthropol.* 32:179–86.

Zuckerman, S. 1932. *The Social Life of Monkeys and Apes*. Kegan, Paul, London.

⌐ Index

Copernicus, Nicolaus, 8
Coprophagous, 117
Cortex: neocortex, 123, 130; visual, 125, 129, 130, 133
Crepuscular, 35
Cynocephalus, 218

Darwin, Charles, 11, 206–7
Daubentoniidae, 33, 96, 104, 106–7, 115, 129, 150, 155, 176, 247
Deciduous (baby or milk) teeth, 102
Deegan, J. F., 132
Deforestation, 247
Dendropithecus, 242
Dental age, 181, 183
Dental comb, 34, 105–6, 203
Dental formula, 102–3, 105, 108, 218, 224, 232
DeVore, Irven, 202
Diastema, 85, 105, 112
Dichromatism, 43
Diet, 114–16
Digestion, 116–21 passim
Disotell, T. R., 71
Distribution, primate, 33–60 passim
Diurnal, 35, 40, 131–32, 198
DNA, 25, 71–74, 243
Dolichocebus, 230
Douc langur. See *Pygathrix*
Drill. See *Mandrillus*
Dryomorphs, 234
Dryopithecus, 236
Dryopithecus (Y-5) dental pattern, 112–13
Duchin, Linda, 165, 209
Dunlap, W. P., 180

Ear: membrane (drum), 85; middle ear, 85, 87; ossicles, 85, 87, 136; external ear, 136–37; Darwin's tubercle, 136
East African Rift System, 235
Egypt, 3–4, 125
Ekgomowechashala, 234
Entepicondylar foramen, 145, 147
Eocene epoch, 216, 220, 221–28
Eosimias, 226–27
Eosimiidae, 226
Erythrocebus, 55, 111, 205
Ethiopia, 115
Euoticus, 38

Euprimates, 222, 224
Europe, 214, 217, 222–25, 233, 236, 243; introduction of monkeys to, 8

Faunivore, 39
Fayum Depression, 224, 226, 230–32
Fleagle, John G., 101, 225, 230
Folivore, 35, 47, 115, 211, 223
Foramen magnum, 77, 88, 94–95
Fossey, Dian, 250
Fouts, Roger, 209
Fowler, James, 246
Frugivore, 36, 55, 107, 115, 211, 223, 231

Galagidae, 33, 35–36, 176
Galago, 36, 38, 128, 166
  G. senegalensis, 119
Galdikas, Birute, 250
Galen, 7–8, 85
Gavan, James A., 31, 185
Gelada baboon. See *Theropithecus*
Gesner, Konrad, 9
Gestation length, 171, 176, 179
Gibbons, 56. See also *Hylobates*
Gibraltar, 53
Gigantopithecus, 237, 239
Gingerich, Philip, 23, 221, 223
Glands: salivary, 116; scent, 204–5
Glaser, D., 135
Golden potto. See *Arctocebos*
Gondwanaland, 214
Goodall, Jane, 61, 209, 250
Goodman, Morris, 24, 72–74
Gorilla, 12–14, 24, 56–60, 97, 167, 179, 203, 210, 241
  G. g. beringei, 11, 15, 58, 196
  G. g. gorilla, 11
Gould, Steven Jay, 180
Graminivore, 115
Grande Coupure ("great cure"), 221, 228
Greece, 5–6, 236
Greek mythology, 7
Gribbin, John, 71
Grooming, 203–4
Growth studies: cross sectional, 184, 192; longitudinal, 184–86, 191